COMPREHENSIVE ENVIRONMENTAL IMPACT ASSESSMENT OF WATER RESOURCES PROJECTS

WITH SPECIAL REFERENCE TO SATHANUR RESERVOIR PROJECT (TAMIL NADU)

COMPREHENSIVE ENVIRONMENTAL IMPACT ASSESSMENT OF WATER RESOURCES PROJECTS

WITH SPECIAL REFERENCE TO SATHANUR RESERVOIR PROJECT (TAMIL NADU)

VOLUME—II

By

Dr. K. B. Chari
Dr. Richa Sharma
Prof. S. A. Abbasi

Centre for Pollution Control & Energy Technology
Pondicherry University
Pondicherry– 605 014

DISCOVERY PUBLISHING HOUSE
NEW DELHI-110002

First Published-2005

ISBN 81-8356-044-X (Set)

© Authors

Published by

DISCOVERY PUBLISHING HOUSE
4831/24, Ansari Road, Prahlad Street,
Darya Ganj, New Delhi-110002 (India)
Phone: 23279245 • Fax: 91-11-23253475
E-mail:dphtemp@indiatimes.com

Printed at:
Amit Enterprises

Dedicated

To

Shri J .P. Singh IPS (Ret)

The Police Chief with a Healing touch

—S. A. Abbasi

My friends—Gajalakshmi, Mallick

And

Ravi—for being with me through the thick and thin

—Chari

Dedication

To

My Parents

—Richa Sharma

Preface

There was a phase during which, in the two decades following the independence of India, creation of reservoirs by the damming of rivers was an activity blessed by the government as well as the public.

But most reservoirs thus created, and the watersheds that fed them, did not happen to be managed in an environment-friendly manner. This wasn't as much due to the inefficiency of the governmental machinery as it was due to the then prevailing lack of environmental consciousness among the executives as well as the lay public. This led to denudation of the catchments, erosion of the watershed soils, silting of the reservoirs, eutrophication, loss of fisheries etc etc. The canal networks in the command areas were also not maintained properly, leading to breeches, weed infestation, puddle formation, and associated evils. Then, again, the available reservoir water was often appropriated inequitably by the more influential farmers to grow too many water-guzzling crops too soon which impoverished the command area croplands leading to diminishing returns. The quicker-than-anticipated rate of siltation of reservoirs meant lesser storage capacity, poorer water quality, and weaker flood control. Most of these problems were caused by improper management rather than any treachery inherent in large dams. But the public ire began to direct itself towards large dams instead of the forces which had contributed to their ill-management.

By the beginning of the 1970s, large dams had acquired the reputation of environment-plundering monsters. This image continued to darken through the 1980s and 1990s during which mini-hydel and micro-hydel projects were increasingly touted as the environment-friendly alternatives to large dams.

But, half-a-decade into the new millennium, we know that mini and micro hydel projects can at best service a small fraction of the requirements of a giant, fast emerging, economic superpower that India is. Further, thousands of dams, constructed at the cost of billions of rupees, are very much in existence. If only we can develop appropriate guidelines to manage these dams in an environmentally sound manner, we can minimize their harmful impact on the environment and maximize the benefits derivable from them. With this objective in view, the Ministry of Water Resources, Government of India, had asked us to conduct an elaborate study of the Sathanur Reservoir Project System (SRPS), situated near Thiruvannamalai, northwestern Tamil Nadu. The present book documents the gist of our findings which aim at assessing the present status of the environmental impact and its management at SRP. In the backdrop of how things are, and what they ought to be, we have tried to develop a body of knowledge that may be helpful in the better management of not only SRP but other dam-based projects as well.

We are grateful to the Ministry of Water Resources, Government of India, for sponsoring the study through its Indian National Committee on Hydrology (INCOH). We were helped immensely by the support and guidance of Dr K. K. S. Bhatia who was the Member-Secretary of INCOH during the tenure of the project. Shri Chetan Pandit who was in that period directing the R&D Division of the Ministry of Water Resources was a consistant source of strength and inspiration. Our grateful thanks to Dr Bhatia and Shri Pandit. We also express thanks for the patronage extended to us by Dr K. D. Sharma, Director, NIH, and for the swift follow-up to our requests by Shri S. Masood Husain (Director & Head, R&D Division, MoWR) and Dr Ramakar Jha (Member-Secretary, INCOH).

S. A. Abbasi

K. Brahmananda Chari

Acknowledgement

We gratefully acknowledge the support of Ministry of Water Resources (Government of India), through its Indian National Committee on Hydrology, for the project which made possible this study.

We fondly thank Prof A. Gnanam, during whose tenure as the Vice Chancellor of Pondicherry University, the project had taken off.

We gratefully acknowledge the help extended to us by the government departments:

Office of the Assistant Executive Engineer
SRBC,
PWD, Sankarapuram
Villupuram

Office of the Assistant Executive Engineer
SRBC,
PWD, Sankarapuram
Sathanur Dam

Office of the Agriculture Officer
Officer of Joint Director
Thiruvannamalai

Office of the Agriculture Officer
Office of the Joint Director
Villupuram

Office of the Assistant Executive Engineer
PWD, Moongilthuraipattu

Office of the District Forest Officer
Kallakurichi
Villupuram

Office of the Director of Public Health Service
Chennai

Office of the District Entomologist
Office of Deputy Director of Public Health Services (DDHS)
Thiruvannamalai

Office of the Divisional Forest Officer
Thiruvannamalai

Office of the Director
Watershed Management Board
PWD, Pollachi

Office of the Director
IHH, Poondi

Office of the Divisional Forest Officer
Kallakurichi
Villupuram

Office of the Directorate of Statistics and Economics
Thiruvannamalai

Office of the Directorate of Statistics and Economics
Villupuram

Deputy Director
Office of the Deputy Director of Health Service
Kallakurichi

Office of the Executive Engineer
Tamil Nadu Electricity Board
Sathanur Dam Hydro-electricity Board
Sathanur Dam

Office of the Executive Engineer
SM and R Division, PWD
Chennai

Office of the Executive Engineer
(Middle Ponnaiyar Region
Thiruvannamalai

Office of the Forest Ranger
Sathanur Dam

Office of the Forest Ranger
Thiruvannamalai

Office of the Health Inspector
D.D.H.S., Kallakurichi

Office of the Health Inspector
Primary Health Centre
Moongilthuraipattu

Office of the Joint Chief Engineer
Water Resources Commission
Tamil Nadu PWD
Chennai

Office of the Medical Officer
Primary Health Centre (PHC)
Sathanur Dam

Office of the Malaria Inspector
Thandarampet

Office of the Manager
Tamil Nadu Fisheries Development Corporation
Sathanur Dam

Office of the Manager
Kallakurichi Cooperative Sugar Mill
Kallakurichi
Villupuram

We thank the support of the following Doctoral, M. Phil, and MCA scholars who helped us in the development of SRPIS:

Rameswaran S, MCA
Development of SRPIS

Girija K
M. Sc, M. Phil
Digital Mapping of SRP maps

Richa Sharma
M. Sc, M. Phil, PhD

Karunakaran N
M. Sc, M. Phil;

Alexander R
MS, M. Phil
SRPIS data collection

Rajalakshmi K
M. Sc, M. Phil
SRPIS Data Organization and Data Entry Coordination

List of Abbreviations Uşed

AIU	Agriculture Impact Unit
API	Annual Parasite Index
EIU	Environmental Impact Unit
IIU	Irrigation Impact Unit
LBC	Left Bank Canal
RBC	Right Bank Canal
SLBC	Sathanur Left Bank Canal
SRBC	Sathanur Right Bank Canal
SRP	Sathanur Reservoir Project
SCA	Sathanur Command Area
SIN	Socio-economic Impact Unit
TRA	Total Reported Area
PHC	Primary Health Centre

Contents

Preface

Acknowledgement

List of Abbreviations Used

Part—B: Development and Application of Geographic Information Systems with Reference to SRPIS (Sathanur Reservoir Project Information System) 581

18. Geographic Information Systems: An Introduction 583

19. SRPIS: An Overview 615

20. Architecture of SRPIS (User's Guide) 617

21. SRPIS Reference Guide 650

22. SRPIS in Future 679

Part—C: Conclusions and Recommendations 681

23. Conclusions and Recommendations 683

Bibliography 709

PART—B

Development and Application of Geographic Information Systems with Reference to SRPIS (Sathanur Reservoir Project Information System)

18

Geographic Information Systems
An Introduction

Among the techniques and tools capable of assisting in resources identification, mapping, and utilization, GIS (Geographic Information System) has had the most spectacular growth. Within a few years of its formal introduction, GIS has come to influence each and every dimension of resource science.

GIS had acquired tremendous power and reach when it was used in tandem with remote sensing. Now, as newer advancements occur in the fields of information technology and communication engineering, the value and impact reach of GIS proportionately increase.

Geographic Information Systems

Geographic Information Systems (GIS) are designed to accept, organize, statistically analyze, and display diverse types of spatial data. These aspects are digitally referenced to a common coordinate system of particular projection and scale. Burrough and Mc Donnell (1998) has categorized various definitions of GIS based on their utility and function:

(a) *Tool Box-based definitions*

(*i*) A powerful set of tools for collecting, storing, retrieving, at will, transforming and displaying spatial data from the real world' (Burrough, 1986).

(ii) 'A system for capturing, storing, checking, manipulating, analyzing and displaying data which are specially referenced to the Earth' (Department of Environment, 1987).

(iii) 'An information technology which stores, analyses, and displays both spatial and non-spatial data' (Parker, 1988).

(b) *Database Definitions*

(i) 'A database system in which most of the data are spatially indexed and upon which a set of procedures operated in order to answer queries about spatial entities in the database' (Smith et. al., 1987).

(ii) 'Any manual or computer based set of procedures used to store and manipulate geographically referenced data' (Aronoff, 1989).

(c) *Organization Based Definitions*

(i) 'An automated set of functions that provides professionals with advanced capabilities for the storage, retrieval, manipulation and display of geographically located data' (Ozemoy et. al., 1981).

(ii) 'An institutional entity, reflecting on organizational structure that integrates technology with a database expertise and continuing financial support over time' (Carter, 1989).

(iii) 'A decision support system involving the integration of spatially reference data in a problem solving environment' (Cowen, 1988).

Geographic Information System (GIS) can also be defined as a set of integrated activities which provides us a tool to:

- Integrate geographic data received from different sources such as maps, charts, tables, aerial photographs, satellite imagery, GPS (Global Positioning System) in digital environment.
- Attach thematic/attribute information to the geographic details.

- Analyze results and build up queries based on spatial and/or attribute information.
- Get the results in a desired form.

The word geographic implies that locations of the data items are known, or can be calculated, in terms of geographic coordinates (latitude, longitude). A Geographic Information System, then according to these criteria may be summarized as having the following characteristics (Martin, 1996):

Geographic: The system is concerned with data relating to geographic scales of measurement, and which are referred by some coordinate system to locations on the surface of the earth. Other types of information system may contain details about location, but here spatial objects and other locations are the very building blocks of the system.

Information: It is possible to use the system to ask questions of the geographic database obtaining information about the geographic world. This represents the extraction of specific and meaningful information from a diverse collection of data, and is only possible way in which the data are organized into a 'model' of the real world.

System: this is the environment, which allows data to be managed and questions to be posed. In the most general sense, a GIS need not be automated (a non-automated example would be a traditional map library), but should be an integrated set of procedures for the input, storage, manipulation and output of geographic information. Such a system is most readily achieved by automated means, and our concern here will be specifically with automated systems.

As suggested by the last point, the data in a GIS are subject to a series of transformations and may often be extracted or manipulated in a very different form to that in which they were collected and entered. This idea of a GIS as a tool for transforming spatial data is consistent with the traditional view of cartography. Thus, GIS have functional capabilities for data capture, input, manipulation, transformation, visualization, combination, query, analysis, modeling and output.

Analog Maps Vs Digital Maps

The conventional maps drawn or printed on paper strive to display information with reference to location. For example, a political map tells us which country, state, or city is located where and a soil map tells us the areas where different types of soils occur. The digital maps in GIS also handle data with reference to space but whereas the conventional maps are static and two dimensional, GIS is dynamic and multifaceted. GIS can not only display information with reference to space but also with reference to time. And GIS can also store, analyze, check, manipulate, and represent data besides displaying it.

Contemporary GIS

Earlier, the GIS development had been focused particularly on the use of networked workstations running under variants of the UNIX operating system. Numerous hardware manufacturers were involved in this field, and the leading software would generally be implemented under the major hardware suppliers systems. Common examples of such machine series at this level were Sun SPARC stations, Silicon Graphics, IBM RS 6000, and Digital DEC stations. Earlier, GIS, because of their heavy use of interactive graphics and database access, would not sit comfortable on mainframe machines running many other tasks simultaneously. Also, GIS have database requirements which until very recently have been too large for most PCs, hence the networked workstation environment was well suited for yester year's GIS applications.

In recent times most of the GIS companies are focusing on the 'desk top GIS' market. With the rapid advancements in the information technology and as well the ever reducing costs of computer hardware, the PC's have become very popular. As the processing power has become more widely available and GIS more widely known, the successively smaller organizations are making use of GIS.

The GIS companies are now trying to grab desktop GIS market by trimming, and redesigning their software into several upgradeable modules. Arguably the most popular of the GIS market leaders—ESRI's—has redesigned and repacked Arc Info, as Arc GIS. The Arc GIS now contains several hierarchical modules:

(i) Arc View - the entry point module into Arc GIS which provides core mapping and GIS functionality (ii) Arc Editor—includes the functionalities of Arc View and bundles some more functionalities such as multi-user editing, versioning, custom feature classes, and dimensioning (iii) Arc Info—the top of in the line of Arc GIS package—can perform GIS data creation, querying, mapping, and analysis.

GIS has the potential to change the geological workplace drastically. As the personal computer has virtually eliminated the typewriter, automated GIS has almost replaced the light table and the map cabinet.

Components of GIS

There are four integrated components of GIS: data and databases, hardware, software including database management systems, and users (Figure—18.1).

1. Data and Database

The data in a GIS are by definition geographic. Spatial data is the information pertaining to where the objects of interest are located. This information can be about the distribution and extent, adjacency, proximity and connectivity, attribute data, or observations about features.

2. Hardware

A fully functional GIS must contain hardware to support data input, output, storage, retrieval, display, and analysis.

Input Devices

(a) *Analog data* such as maps, satellite images and photographs need to be converted to *digital data,* in raster format or vector format, for which a scanner or/and digitizer is required.

(i) *Scanner.* It is an electronic optical device that converts analog data such as maps drawn on a paper into the 'raster' format. A *raster* is a type of computerized picture logically made-up of a two-dimensional array of cells, like a spreadsheet. Aerial photographs and satellite imagery are common types of raster data used

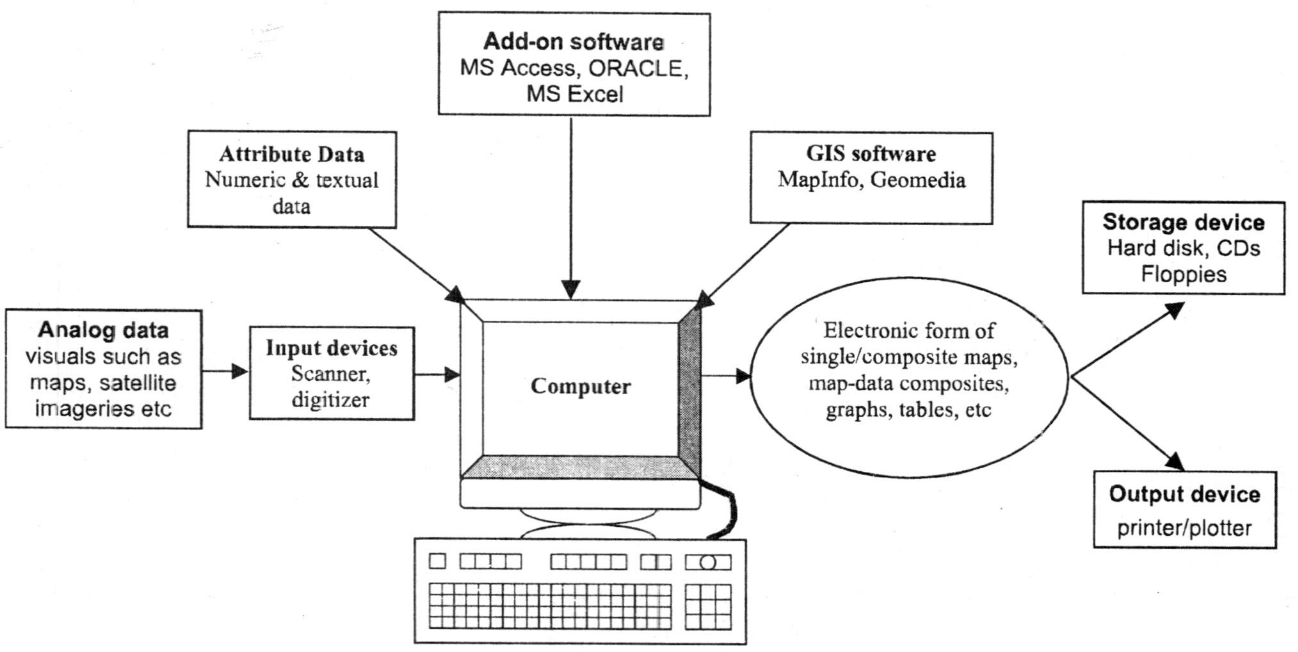

Fig. 18.1: Components of GIS

in GIS. Some of the most familiar raster formats are TIF (Tagged Image File). BMP (Bitmap), and GIF (Graphics Interchange Format). A composite colour raster uses a colour table to map each raster cell value to a discrete display colour.

(ii) *Digitizer:* A device that enables converting any conventional paper map into such an electronic form that has specific position (in the form of coordinates) associated with each bit of the image. The digitizer consists of a table (or tablet) onto which a paper map is attached. The map is then traced by moving a hand-held, mouse-like device known as a cursor (also known as puck/transducer), across the surface. This results in the conversion of bits of the map into corresponding electronic version with coordinates associated with each bit.

A Vector is a co-ordinate based data structure commonly used to represent map features as point, line and polyline. Each object is represented with reference to x and y co-ordinates.

Some of the newer GIS software—MapInfo and Geo Media among others contain modules with which scanned images can be digitized onscreen without the digitizer hardware. As the onscreen digitizing provides features that are much more powerful and cost effective than the hardware-based digitizer, the later is becoming increasingly redundant.

(b) *Computer:* Computer forms the core hardware which stores and processes all information. Aside GIS software, a computer can also house statistical, graphical, and animation software which, depending on the user-friendliness of the GIS software, can be interfaced with the latter. The output can accordingly be enhanced in terms of quantity as well as quality.

(c) *Storage devices:* The maps and/or databases developed by the computer (with the help of GIS and other software) can be stored on digital media such as Compact Disc (CD), and floppy disc.

(d) *Output devices:* Printers and plotters constitute the output devices. Even a black-and-white dot-matrix printer can transfer a GIS map from its electronic form to a paper but to achieve distinct representation of various features in a map - with adequate tonal quality and contrast-colour inkjet/ LaserJet printers of 1200 dpi or better resolution are needed.

3. Software

Some of the most popular GIS software are Arc Info, Arc GIS, MapInfo, Geo Media and TNT-MIPS. Each of these software offers different levels of functionality. It is in the interest of the GIS user to perform an assessment of the GIS requirements prior to committing purchase of the GIS software.

4. Users

Today, GIS is used by diverse professionals—from a cartographer to a commercial pizza dealer! Among the disciplines such as geology, hydrology and social sciences GIS has become as indispensable a tool as word processing and spread sheet software such as MS Excel.

To make the most of a GIS the user however need to be well versed in aspects such as map reading, database management, spatial analysis, computer cartography, computer science, programming, and basic geography.

Technical Elements of a Digital GIS

A GIS is built around a framework of five basic technical elements: (1) encoding (2) data input (3) data management (4) manipulative operations and (5) output products (Figure 18.2).

Encoding

GIS mapping essentially deals with three distinct spatial attributes: points, lines and polygons. These spatial entities can be encoded by using two different types of position indexing systems: (1) grid-cell or raster coding and (2) polygon or vector coding. Grid-cell coding is conceptually a matrix system superimposed over the geography such that the attribute information can be collected by a systematic array of grid squares or cells. Normally,

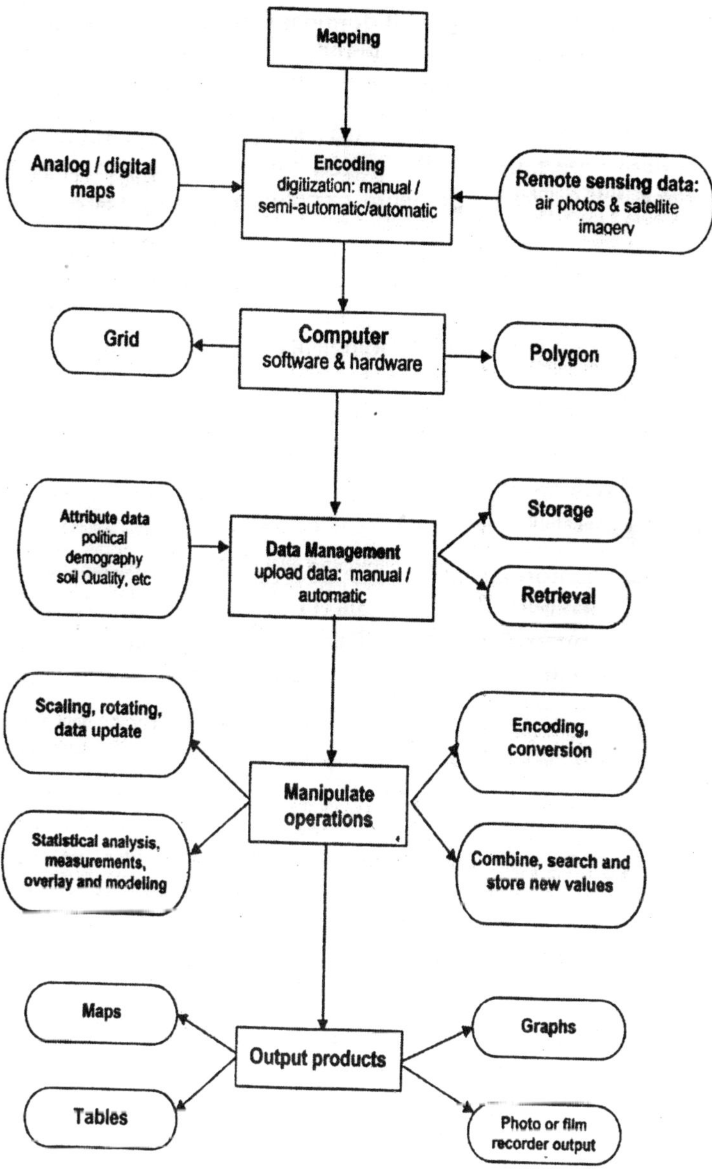

Fig. 18.2: Various steps involved in a typical GIS

the information category most dominant for each cell is encoded. As the cell size largely determines class accuracy, two methods have emerged to assist in preserving data integrity:

(a) decreasing cell size and (b) listing the relative amounts of each data type falling within a cell. Grid cells are functionally identical to the picture elements, or pixels, that compose a digital image.

In polygon coding, the perimeter of each unit area containing the desired attribute data is digitally encoded and stored. One type of polygon indexing is topological coding, whereby areas are formed by connecting polygons, and polygons are formed by connecting arcs. Polygon coding defines bound areas more accurately and requires less computer storage space than does the grid-coding structure.

Data Input

Analog information (e.g. maps on paper) is converted to the digital domain by the process called digitization. This forms an essential component of GIS. There are several methods for digitization: manual, semi-automatic and automatic.

Manual digitization involves tracing the analog maps using a specific hardware called table digitizer and other accessories such as puck / transducer. Semi-automatic digitization involves the use of a scanner, GIS software and the mouse of the PC. The analog maps after scanning (in the raster form) are registered using the GIS software. Once these scanned maps are registered, various topological features of these maps are redrawn and saved as several layers such as vegetation, settlements etc.

The automatic digitizing involves hardware (a scanner) and the GIS software. The scanned maps or the raster images can be automatically digitized using some of the GIS software at one go. Later, the user would have to sieve the various digitized features, then select, cut and paste into various layers.

Data already in the digital form (e.g. satellite images) usually have to be reformatted and scaled to match the geometry of the GIS reference map projection.

Data Management

Data management is extremely important for successful and efficient operations of a GIS. Because of the large volume and the variety of data available these days, plus the wide range of potential applications. Data management consists of series of computer programmes to perform all data entry, storage-retrieval and maintenance tasks (Figure—18.2)

Manipulative Operations

GIS are capable of performing two kinds of automated analysis, surface analysis and overlay analysis. Surface analysis applies to intra-variable relationships that exist within one data plane. For example, soil categories can be grouped together, analyzed and labeled according to agricultural value. Most surface analysis produces new variables that can be applied to other surface or overlay analysis procedures.

Overlay analysis, as mentioned earlier, involves a set of distinctive map layers, the various features of the map stored in such a way that the user can view the layers of his choice.

Output Products

A GIS can retrieve and display data in graphic (as maps, bar charts, line graphs) or tabular form, or both. Most systems are capable of producing hardcopy charts, scatter diagrams, tables, and maps in various forms and sizes. In addition, all systems have a PC monitor on which graphic or tabular information for segments of a data base or multiple data base can be displayed. This represents interactive analysis because the retrieval and display of data are in near real time (Avery and Berline 1992).

Basic Functions of a GIS

User Interface

This is an important aspect of any software. It is necessary to choose software which provides a comprehensive set of capabilities, designed in such a way that the user can get used to the software in a very short time.

Some of the standard user interface of the GIS software are: wizard-like dialog boxes, short cut menus, status bar pop-ups, flexible toolbar positioning by dragging and the linked views of the map window and the browser window (Figures—18.3 and 18.4).

Mapping

As mentioned earlier, GIS enables dynamic mapping of the geography-earth and its features. Some of the features that make GIS dynamic are: layer control, legend display, zooming, and panning of the maps.

Layer control. This feature helps in overlaying several map layers and control them—which layer to be displayed and in which order.

Layer intelligence. When several map layers are opened, the GIS software orders the layers appropriately, i.e., points on top, boundaries underneath. This saves a lot of time by not requiring to reorder the maps while adding them.

Legend. Some software such as Arc View enables automatic display of map legends and in some software, such as MapInfo, it can displayed by choosing the menu.

Zooming, panning and map auto scroll features add convenience by keeping the most important area visible.

Built-in-hot link capability enables one to display text files, images, other components of the project by clicking on the map object.

Data Analysis

The queries in GIS can be of geographic in nature, text based or through ODBC (Ole Data Base Connectivity). Queries in GIS can be expressed as avenue expressions or by SQL (Structured Query Language). Presently SQL based queries have gained much popularity among the users.

Desktop mapping tools enable the user to ask questions and get answers from the data. This helps in understanding the geographical relationships hidden in the database.

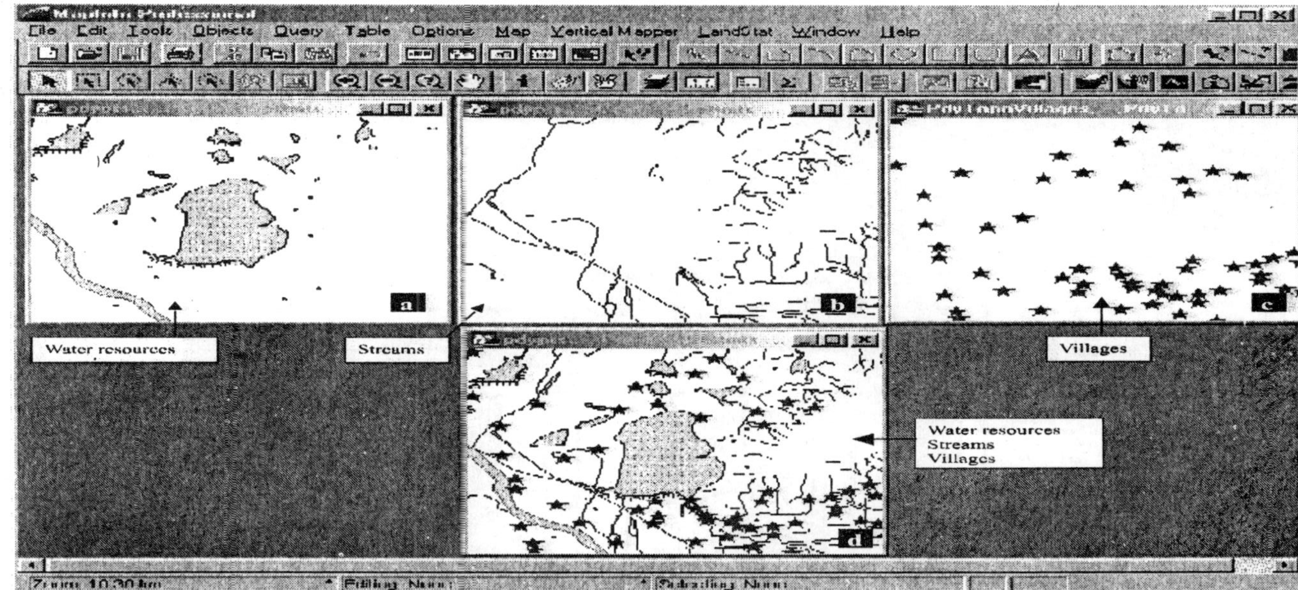

Fig. 18.3: **The building blocks of a typical GIS: the layer 'a' features the polygon objects (in this case water bodies), layer 'b' features polylines (water streams) and the layer 'c' features point objects (villages). An overlay of the above three layers makes a complete map (d).**

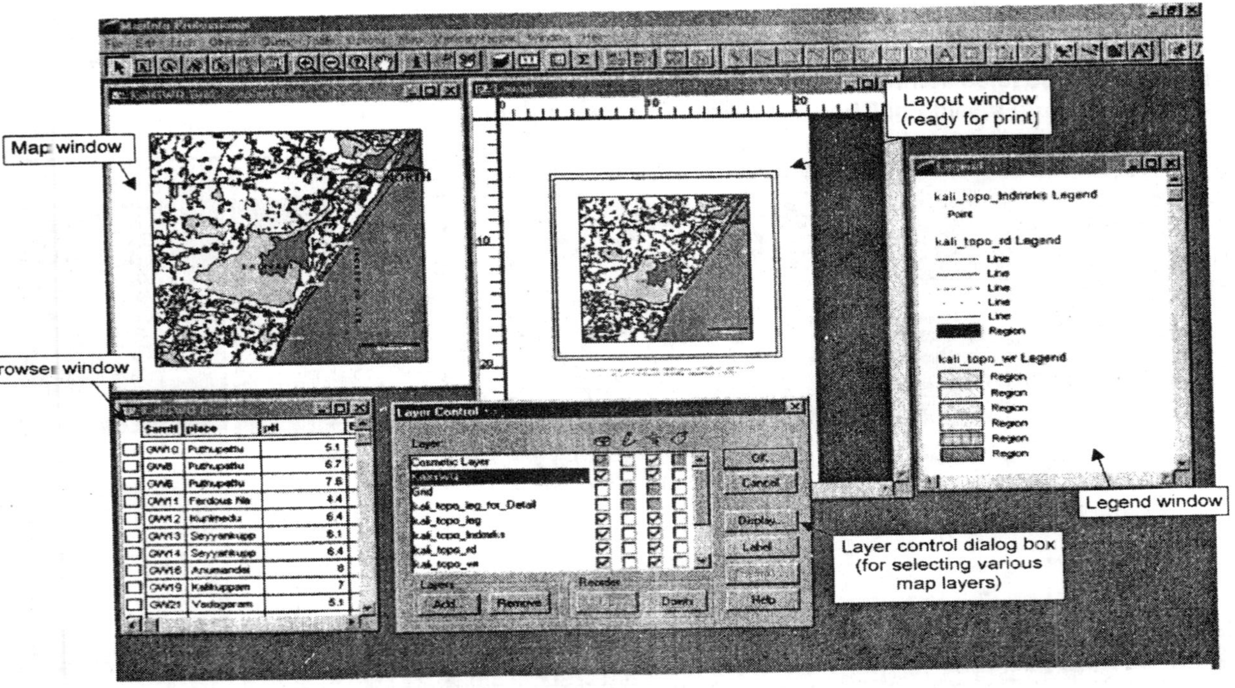

Fig. 18.4: A typical interface of GIS software displaying various features

Information retrieval by clicking on the map. Usually the Info tools show the associated attributes of the map. The Button tools enable to select map objects or features by their proximity to other map objects.

SQL support. The query builder incorporates geographical queries and functions. This includes geographic operators and functions in the relational SQL queries. This feature is highly useful in performing queries on multiple layers, for example one can select all the point objects that are present in the polygons of some other layer and vise versa.

Geographic operators. This feature provides the power of dynamic mapping. It enables the user to understand the geographical relationships within the database. Some of the most widely used geographic operators are: (i) completely within and completely contain (ii) have their center in, contain the centre of (iii) intersect (iv) are within the distance of.

Redistricting. This feature, also known as load balancing, allows the user to create new districts and realign existing districts, while doing calculations of the attached data for instant decision making. For example a manager can easily balance sales territories by the number of customers in the territories.

Find tool. A feature on the map or a record in the database can be easily searched using the *find tool.* This requires that the particular field is indexed, which is being searched.

Geographic Data Manipulation

Geographic data manipulation allows the user to create and edit geographic data and associate tabular information so that they can be analyzed more quickly and easily.

Polygon overlay capabilities with data aggregation and disaggregating makes it possible to combine, split, and erase map objects. This would make it easier to update the maps and their associated data .

Some of the popular tools for geographic object editing, which make digitizing and mapping easier are: overlaying of nodes at intersections, auto-tracing of polylines and polygons, convert polylines and polygons, snapping to the nearest nodes and auto-completing of polygons.

Buffer tool makes it easier to perform geographical queries and analysis. This feature is particularly helpful when performing multiple what-if scenarios. The data driven buffers, which use the underlying database information to determine buffer size, increases the flexibility in buffer creation and gives more accurate and / or more content rich means for visualizing the data. For example, for a layer of points that represent radio towers, one can create individual buffer zones based on each towers broadcast power.

Projection and Coordinate Systems. Support for various geographic projections is necessary when working with international data sets. Some of the standard features include: displaying of maps that are stored in numerous projections, supporting projection conversion, and supporting custom coordinate systems enables in digitizing, correctly overlaying maps, and to permanently change and save a map into a new projection.

Data Access

Large volumes of data are collected in different formats such as spreadsheets (MS Excel, lotus 1-2-3), delimited ASCII files and relational databases (ORACLE, MS Access). Accessing these data in a fast and easy way with the desk top GIS would enhance the productivity and usability of a software.

Usually the spreadsheet data is accessed on the fly, while RDBMS (Relational Database Base Management system) can be accessed using the ODBC connectivity. Having a remote SQL read/ write feature would not only save a lot of the time but makes it not necessary to store the intermediate copies of data locally.

Spatial Data (Map Files)

Compatibility. With a wide range of GIS software now catering to the requirements of GIS users, it is essential that the map data / files developed in one GIS software are readable / editable in other GIS software. Most of the GIS software support popular formats.

Universal translator. This feature enables the user to translate file formats of one software to the other required format.

Raster image support. This feature would enable to register raster images to the world coordinates and overlay these with other vector formatted data.

Spatial database support. Read / write support for SpatialWare, ORACLE, Autometrics, etc would allow to store geographic (map) data in RDBMS. This open architectural approach to client / server implementation of geographic queries eliminates re-geocoding of the data.

The compliance of database for point data (coordinates, appropriate spatial indices, etc) enables remote geographic queries to be performed on the server.

Spatial Statistics

The dominant feature of GIS technology is that spatial information is represented numerically, unlike in the case of analog paper maps. The analog nature of map sheets, manual analytic techniques are principally limited to qualitative processing. Digital representation on the other hand, has the potential for qualitative as well as quantitative processing.

The ever growing field of GIS has stimulated the development of spatial statistics, a discipline that seeks to characterize the geographic distribution or pattern of mapped data. Spatial statistics differs from traditional statistics by describing the more refined spatial variation in the data, rather than producing typical responses assumed to be uniformly distributed in space.

Map Algebra

Just as spatial statistics has been developed by extending concepts of conventional statistics, a spatial mathematics has evolved. It is similar to traditional algebra, in which primitive operations (add, subtract, exponentiation) are logically sequenced on variables to form equations, but in map algebra, entire maps represent variables. The logical sequence involves retrieval of one or more maps from the database, processing those data as specified by the user, creation of a new map containing the processing results, and the storage of the new map for subsequent processing.

The cyclical processing is similar to "evaluating nested parenthetical" in traditional algebra. Values for the "known" variables are first defined, then they are manipulated by performing the primitive operations on those numbers in the order prescribed by the equation. For example, in the equation A = (B+C)/D the variables B and C are first defined and then added, with the sum stored as an intermediate solution. This intermediate value, in turn, is retrieved and divided by the variable D to derive the value of the unknown maps. The numbers contained in a solution map (in effect, solving for A) are a function of the input maps and the primitive operations performed. In a similar manner per cent change in any two maps can be calculated.

Visualization and Presentation

Visualization helps in understanding data through its visual representation. The end result of most of the GIS packages must have tools to make quality hardcopy or softcopy output.

A robust GIS software wouid provide multiple views of the data in the form of maps, charts, tabular views (rows and columns), and the ability to combine all these views (Figure—18.4).

Maps—feature styles and labels. A good GIS software provides numerous style choices: pen styles (line styles, thickness, how end points meet), brush styles (fill patterns, including transparent fills), symbols (including custom symbols).

Maps—thematic representation. Thematic maps enable representation of data on a map with various colors, fill pattern, line styles, and symbols. Some of the thematic map types include: ranged maps (graduated by colour and size); individual (unique value); chart maps (pies and bars), dot density, and custom types); graduated symbol (the size of the symbol is proportionate to the data values of the points); bivariate or multivariate thematic analysis; and grid surface map (provides continuous colour gradation across the map).

These thematic map types enable the user to represent more than one variable in one map window. For example one can represent income by the size of the symbol and show education level by the symbol's colour. Also, it is possible to overlay more than one thematic map in a single map window. A polygon thematic map can be overlaid by the point based thematic map layer.

Charts and graphs. The GIS software also enables to create area, line, bar, pie and scatter plots with flexible display options (rotated, 3-D, etc) allowing to represent the data with more traditional business visualization tools without relying on other software applications.

Presentation. The final layout map with the WYSIWYG (What You See Is What You Get) standards enable one to make the final output ready for presentation.

Scalable Mapping Solutions

Scalable mapping solutions enable the user to leverage the power of geographic relationships regardless of the choice of application or implementation.

Scalable hardware requirements. This feature allows one to adjust the hardware configuration based on the organization requirements. For the software which have provision of hardware lock can readily be upgraded with higher end RAM, HDD etc. However, in the case of floating/node locked software, which depend on the HDD identity or operating system or RAM or a combination of these, one would require to renew the license.

Development environment. A development environment allows one to: (a) extend the functionality of the base software (b) customize the user interface (c) automate repetitive processes and (d) integrate the basic software with other applications.

Role of GIS in the Environmental Management

GIS can be very useful in (i) bringing forth the hidden patterns in a dataset (ii) perform queries (iii) store, edit and retrieve data in the form of maps, tables or graphs and (iv) prepare exceedingly 'expressive' maps for publishing In conjunction with statistical modelling, mathematical modelling, and the techniques and tools of operations research, GIS is becoming an increasingly potent device for facilitating environmental management.

Some aspects of the role of GIS in environmental management are mentioned below.

Delineation of Land Use and Land Cover

Land use and land cover of a watershed /river basin can affect the quality of surface / ground water directly. For instance, if the presence of land cover in the form forests dominate in a catchment it would reduce the possibility of soil erosion. In case agriculture happens to be the most dominant land use, it would contribute agricultural inputs in the form of fertilizers and pesticides, and soil sediments to the water resources as run-off or through infiltration.

Overlay Analysis

This feature of GIS can be highly useful to prepare a new map by overlaying several maps. For example a map of soil erosion index can be prepared by overlaying individual maps of digital elevation model (or three-dimensional elevation, which gives information on the slope of the terrain), soil type, and land cover (which provide clue to the potential of runoff; Figure—18.5).

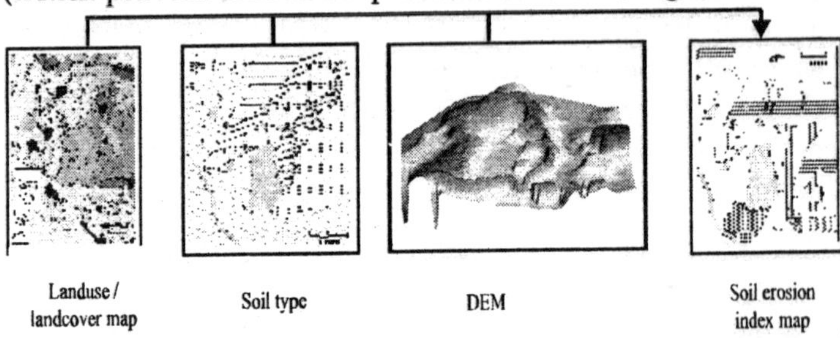

Fig. 18.5 Deriving a soil erosion index map using overlay analysis

Buffering

This feature can be used for delineating the risk zones or a new region around a known feature. For instance if it is known that regions which are within 2 km of a reservoir would be flooded, these regions can be delineated using the buffer tool (Figure—18.6).

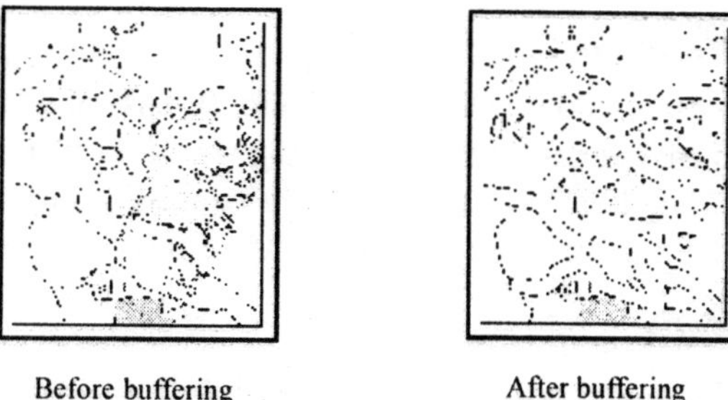

Before buffering After buffering

Fig. 18.6: A drainage pattern *before buffering* and *after buffering*

Aggregation or Redistricting

This feature is useful in the mapping and performing of what if analysis. Say, if there are three classes of vegetation which are delineated—jungle, scrub jungle, and open scrub. The user wishes to club these into two groups—jungle and scrubs. The user can do this easily by clubbing the scrub jungles and open scrub, and labeling this new group scrub by using the feature called redistricting.

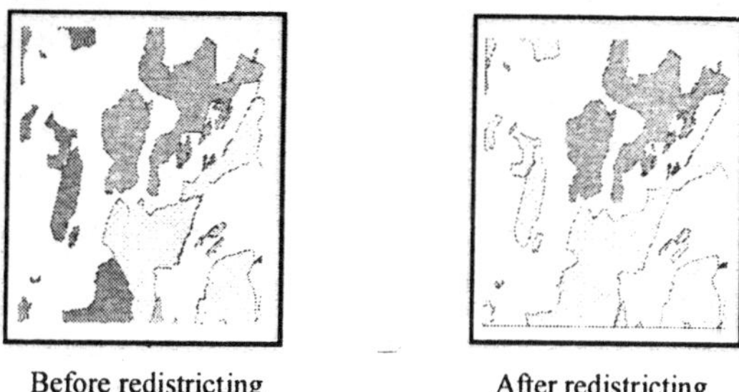

Before redistricting After redistricting

Fig. 18.7: Three groups of land cover converted to two groups using the *redistricting feature*

Network Analysis

This feature enables the user to find out the shortest path between two locations, say the distance between one sampling site and the other. The pattern and the distance a stream flows in a drainage basin can also be calculated by using network analysis.

Viewshed Mapping

A three dimensional elevation model (DEM) of a landscape or the region of interest enables viewing the terrain in different perspectives. The DEM can also provide vital information on the terrain slope, which can be useful in arriving at soil erosion index.

Thematic Maps

Thematic maps represent numerical data in graphic form. Thematic maps have a better visual appeal, and are easy to understand, than the tabled data and graphs (Figure—18.8).

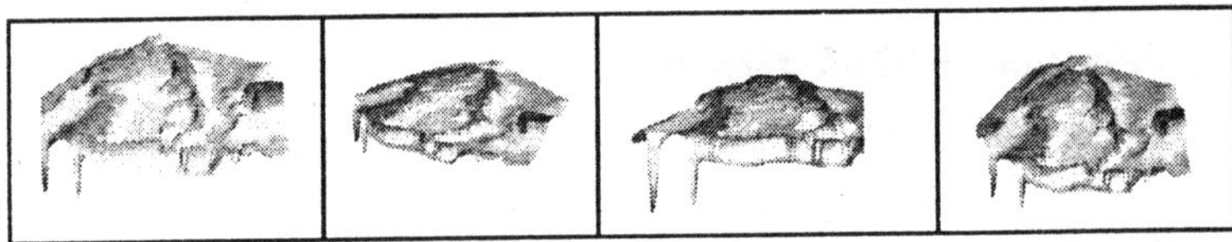

Fig. 18.8: Different perspectives of a 3D model

Spatial Analysis

The spatial data can be intrapolated or extrapolated using several models such as nearest neighborhood and krigging. The spatial data can also be represented as line contours (elevation contours as lines) or as a regions (elevation contours as regions), or as grid (a map representing various data values as continuous gradient of colours). Further, the grid can be represented as an 3 D elevation model (Figure—18.9).

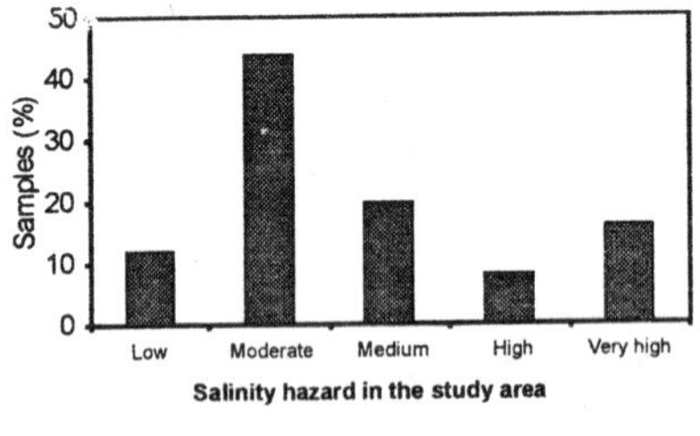

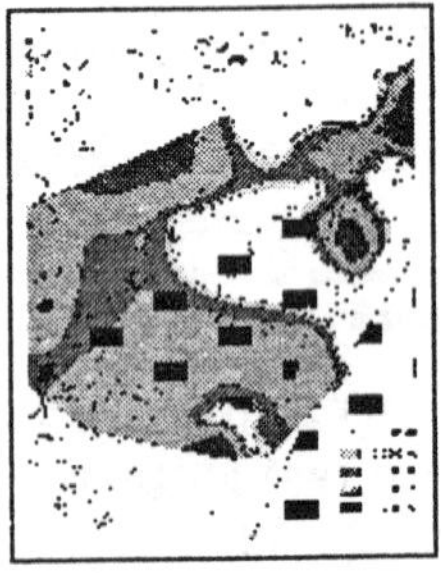

Fig. 18.9: Spatial data represented as a *graph* and a *thematic map*

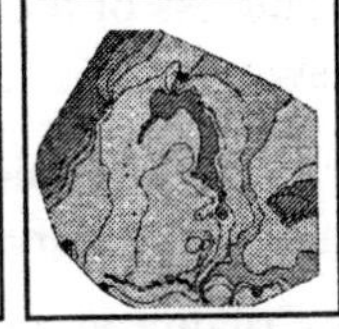

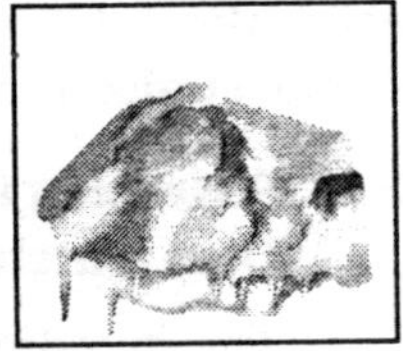

Fig. 18.10: Various forms of representing spatial data

Application of GIS and Remote Sensing in Water Resources Management

We present below a few illustrative examples of the successful application of GIS and Remote Sensing in water resources management.

1. Impact of human interference on Kaliveli Wetland (Chari and Abbasi 2002, 2003; Chari 2002)

Kaliveli, about 130 Km south of Chennai and close to Pondicherry is one of the largest wetlands of Asia. It is a very important wintering ground for migratory birds and has been identified as a heritage cite by the International Union for the Conservation of Nature. In order to develop a system which could continuously assess the impact of human activities on Kaliveli wetland, Chari and Abbasi (2002, 2003), have developed a GIS by integrating topo sheet maps, satellite imageries, and ground truth studies with the help of software tool MapInfo Professional 5.5. The resulting GI system not only helps in identifying impacts but also throws up suggestions on how to mange the wetland in an integrated, sustainable manner.

2. Monitoring Small Dams in Semi-arid Regions Using Remote Sensing and GIS

The analysis of data from high spatial resolution satellite-mounted infrared sensors has the potential to monitor the water stored by small dams in semi-arid areas. Finch (1997) has described a simple method of analysis, giving particular attention to selecting an appropriate threshold level to identify land with or without water cover. The use of a GIS allowed further delineation of areas with water cover from other areas, such as deep shadow, which the sensor had earlier classified incorrectly, to facilitate automatic calculation of dam capacities. The success of the technique was demonstrated in two regions of Botswana.

3. Assessing the Vulnerability to Soil Erosion of the Ukai Dam Catchments Using Remote Sensing and GIS

The investigation of basins for planning soil conservation requires a selective approach to identify smaller hydrological units, which would be suitable for more efficient and targeted conservation management programmes. One criterion, generally used to determine the vulnerability of catchments to erosion, is the sediment yield of a basin. In India, sediment yield data are generally not collected for smaller sub-catchments and it becomes difficult to identify the most vulnerable areas for erosion that can be treated on a priority basis. An index-based approach, based on the surface factors mainly responsible for soil erosion, is suggested in a study by Jain and Goel (2002). These factors include soil type, vegetation, slope and various catchment properties such as drainage density, form factor, etc.

The method adopted by Jain and Goel (2002) has been illustrated with a case study of sub-catchments immediately upstream of the Ukai Reservoir located on the River Tapi in Gujarat State, India. The area is divided into 16 watersheds and different soil, vegetation, topography and morphology-related parameters are estimated separately for each watershed. Satellite data are used to evaluate the soil and vegetation indices, while a GIS system is used to evaluate the topography and morphology-related indices. The integrated effect of all the parameters is evaluated to find different areas vulnerable to soil erosion. Two watersheds were identified,as being most susceptible to soil erosion. Based on the integrated index, a priority rating of the watersheds for soil conservation planning is recommended.

4. Geographic Information Systems (GIS)-based Spatially Distributed Model for Runoff Routing

GIS offer many new opportunities for hydrological modelling. They can be used to form spatially distributed models of watershed. However, some problems of this approach, e.g. the parameterization of physically based models, are not resolved yet. Conceptual models of the meso-scale still have a great practical importance.

Hrissanthou et. al., (2003) have used a rainfall-runoff sub-model for the computation of the surface water volume inflowing into the lake from the main streams of the sub-basins located around Kastoria Lake. A quasi-three-dimensional simulation model of the Kastoria basin aquifer is also realized, in order to estimate the groundwater contribution to the volumetric budget of the lake and the whole basin as well. For the computation of sediment load inflowing into the lake from the main streams of the sub-basins, the rainfall-runoff sub-model is combined with a soil erosion sub-model and a sediment transport sub-model for streams. A GIS was developed in the hydrologic basin with all data needed for parameter identification and model application. The data base was enriched by a series of on site measurements of water discharge made in all main streams for one whole hydrologic year. By means of the resulting mathematical sediment model, those sub-basins, which deliver most sediment load to the lake, are identified. On the basis of this identification, a series of control measures, for the reduction of sediment inflowing into the lake, at certain places of the above mentioned sub-basins is proposed.

Schumanna et. al (2000) have presented an approach: how statistical descriptions of distributed catchment characteristics could be used to consider spatial heterogeneity within conceptual models. Three semi-distributed modules are presented. The three components are combined to a hydrological model including feedback components between surface flow and infiltration and between subsurface return flow and surface flow in saturated areas. The model was set up to use spatially distributed information about catchment characteristics for the estimation of its parameters. By a direct estimation of some model parameters from a GIS-based analysis of the catchment characteristics, the number of calibration parameters can be reduced. In the second part it is shown how the application of this model to different catchments within a region can benefit from boundary conditions for optimization, which are derived from a GIS considering the differences of catchment characteristics.

Olivera and Maidment (1999) have proposed a method for routing spatially distributed excess precipitation over a watershed to produce runoff at its outlet. The land surface is represented by a (raster) digital elevation model from which the stream network is derived. A routing response function is defined for each digital elevation model cell so that water movement from cell to cell can be convolved to give a response function along a flow path and responses from all cells can be summed to give the outlet hydrograph.

5. Changes in the Water Quality Indicated by Submerged Macrophytes Using GIS

Geographic Information Systems (GIS) are useful for mapping and storing information on submerged vegetation allowing easy interrogation, updating and plotting of spatial information at various scales, and providing a reference for future comparisons.

Lehmann and Lachavanne (1999) have compared the distribution of submerged macrophytes along 20 km stretch of lake shore (Lake Geneva, Switzerland) between the years 1972, 1984 and 1995. Lake Geneva underwent rapid eutrophication until 1980, followed by a reversal that is still in progress. *Potamogeton pectinatus, P. perfoliatus, P. lucens* and *Elodea Canadensis* showed no significant changes in their distribution, with the two former species dominant throughout. *Chara sp.*, and to a lesser degree *Myriophyllum spicatum*, decreased in abundance between 1972 and 1984 but had increased again by 1995. The abundance of *P. pusillus* increased regularly, while *Zannichellia palustris* and P. crispus almost disappeared from the study area. *Elodea nuttallii* was observed for the first time in Lake Geneva in 1995.

Two methods of bioindication of water quality by macrophytes are compared. The macrophyte index proposed by Melzer (1988) is based on nutrient load, whereas the saprobic index proposed by Sladecek (1973) measures organic pollution. The saprobic index is sensitive to small changes in species composition and abundance, and also reflects better the changes in eutrophication. It may therefore be a better bioindicator.

6. Integrated Geographical Assessment of Environmental Condition in Water Catchments

Water catchments are functional geographical areas that integrate a variety of environmental processes and human impacts on landscapes. Integrated assessments recognize this interdependence of resources and components making up water catchments and are vital for viable long-term natural resources management. The work by Aspinall and Pearson (2000) couples eco-hydrological modelling with remote sensing, landscape ecological analyses and GIS to develop a series of indicators of water catchment health as part of a geographical audit of environmentai health and change at regional scales.

Indicators are simple measures that represent key components of the system and have meaning beyond the attributes that are directly measured. A suite of indicators, many capable of measurement from remote sensing data sources, are described that represent state (condition) and trend (changes across space and time) and focus on the physical, biological and chemical properties of water catchments, as well as their ecological function (stability, resilience, and sensitivity). Models implemented in GIS allow indicators to be combined within water catchments by setting them within a specific geographic context and integrating the descriptions of environmental variability across the geographic area. This spatial integration is necessary to place individual, site-specific indicators within a broader geographic context; the models allow this context to reflect the ecological and hydrological functioning of the water catchment. Scale and other geographic effects associated with integration are managed using an approach that partitions the landscape into a hierarchical series of nested functional units.

Methods from image analysis, landscape ecological analysis, spatial interpolation, and numerical process modelling are integrated within a GIS (Arc View) to provide a single environment within which to conduct the study. Results are described from the catchment of the upper Yellowstone River in the Rocky Mountains, USA, an area of about 14000 km^2. The river source is in Yellowstone National Park. The catchment is subject to a number of land-use issues notably those associated with changing patterns and types of land use including forestry, irrigated agriculture, range management, wildfire, mining, summer and winter recreation, and residential development, which are associated with a number of land-use conflicts and impacts.

7. Application of GIS in Fisheries Science

Introduction of GIS in fisheries science has been relatively recent compared to other disciplines. This is due, in part, to the inherent complexity of the environment where fish life cycles take place and to the complexity of fisheries dynamics as well. Geographic information systems have been developed mainly for terrestrial landscapes and many of their algorithms are not suitable for marine ecosystems. The principal challenges in the implementation of geographic information systems to fisheries research have to do with the three dimensional space where fish live, temporal variability and fuzziness of the data sets. There are still few studies in fisheries science which incorporate GIS in their research. In a research article Malavear (2002) has reviewed some of the areas in fisheries research which are currently using geographic information systems, the positive outcomes of the integration of GIS in fisheries science, enumerate some methodological problems found and how these studies deal with them and finally outline some potential areas for further research.

Marrsa et. al (2002) have done a 4-month pilot study on micro-scale mapping of Nehrops trawler effort in the Clyde Sea area, west Scotland, using global positioning system (GPS)-based position data loggers. The position data loggers have provided unbiased, accurate fishing effort data at a scale hitherto unrecorded. The trawl location data were combined with the daily landings obtained from a complementary confidential logbook scheme using GIS to produce maps of fishing effort, cumulative landings and landings per unit effort.

In an another interesting and promising application of GIS in fisheries management, Schultz et. al (2002) have studied the crayfish occurrence in relation to land use properties using GIS. The authors have used spatial analysis to identify the influence of land use properties and other human impact on crayfish distribution at landscape level.

8. Using GIS to Assess Environmental Risk

Surveillance of Mosquito Vector Habitats and Risk Assessment

Dale et al (1998) have successfully used remote sensing to achieve multiple objectives focusing on mosquito management. The authors have demonstrated how Geographic Information Systems, combined with remote sensing analysis, have the potential to assist in minimizing disease risk.

Mapping the breeding habitats of the species, *Culex annulirostris* and *Aedes vigilax*, facilitates assessment of the risk of contracting the diseases and also assists in control of the vectors. Colour infrared aerial photography was used to identify the specific parts of the salt marsh in which larvae and eggs of Ae. vigilax are found. A simple risk model has been applied to the field data and then linked to computer-aided analysis of remotely sensed data to map potential ephemeral breeding sites. This application has the potential to guide control at critical times, for example after heavy summer rainfall or when there is an outbreak of Ross River virus disease.

The authors have explored novel ways to map the detailed pattern of water under mangrove forest canopy to identify where mosquitoes are breeding and as an aid to planning modification.

A Watershed-based Cumulative Risk Impact Analysis: Environmental Vulnerability and Impact Criteria

A watershed-based screening tool, the Cumulative Risk Index Analysis (CRIA), was developed by Osowski et. al (2001) to assess the cumulative impacts of multiple CAFO (Swine Concentrated Animal Feeding Operations) facilities in a watershed subunit.

The CRIA formula calculates an index number based on: 1) the area of one or more facilities compared to the area of the watershed subunit, 2) the average of the environmental vulnerability criteria, and 3) the average of the industry-specific impact criteria. Each vulnerability or impact criterion is ranked on a 1 to 5 scale, with a low rank indicating low environmental vulnerability or impact and a high rank indicating high environmental vulnerability or impact. The individual criterion ranks, as well as the total CRIA score, can be used to focus the environmental analysis and facilitate discussions with industry, public, and other stakeholders in the agency decision-making process.

Flood Risk Mapping

Geographic Information Systems (GIS) are an efficient and interactive spatial decision support tool for flood risk analysis. The paper by Sinnakaudan et. al (2003) describes the development of ArcView GIS extension—namely AVHEC-6.avx—to integrate the HEC-6 hydraulic model within GIS environment. It has the capability of analyzing the computed water surface profiles generated from HEC-6 model and producing a related flood map for the Pari River in the ArcView GIS. The user-friendly menu interface guides the user to understand, visualize, build query, conduct repetitious and multiple analytical tasks with HEC-6 outputs.

The flood risk model was tested using the hydraulic and hydrological data from the Pari River catchment area and the results of this study clearly show that GIS provides an effective environment for flood risk analysis and mapping.

9. Landscape Visualization Techniques Using GIS

Visual representations are increasingly used to communicate the impacts of environmental changes. Geographic information systems (GISs) are becoming common sources of the spatially organized data needed to create valid and defensible visualizations. However, these data lack the detail and richness needed to create the realistic imagery felt to be critical for public review. The paper by Orland (1994) describes exploratory studies in the application of techniques drawn from remote sensing and applied to ground-level photographic images of sensitive locations to achieve realistic images with demonstrable relationships to an underlying GIS. Digital filtering and image sampling processes have been used to simulate the visual consequences of forest pests, of timber management activities, of forest wildfires, and of recovery from all of these impacts. The resulting images have been used to communicate expected outcomes to participants in policy-development settings, and to initiate the development of public perception models relating impacts to public preferences. Although integration of these techniques with GIS systems has not yet been achieved, the necessary development steps are outlined by the authors.

Danahy (2001) suggests that the visual media commonly used to structure scientific analysis, professional design, decision-making and artistic interpretation of visual landscapes are quite weak at portraying the dynamic and peripheral dimensions of human vision. The absence of a convenient, cost-effective means for showing all the fundamental visual aspects of landscape in a balanced way is a serious limitation. This paper by the author outlines this particular issue and proposes that as electronic media and computational media become more developed and are applied to the realm of visual concerns, it will become more practical to include peripheral vision and dynamic viewing in deliberations about visual landscapes. Also, the paper reflects on the potential of visualization automation techniques to overcome these shortcomings through illustrations of project work using innovative software tools developed to explore this question at the Centre for Landscape Research (CLR) at the university of Toronto.

10. GIS Based Decision Support Systems

A Web-based Environmental Decision Support System (WEDSS) for Local Government Planning

In an under publication report Sugumaran et. al. (2004), describe the development of a Web-Based Environmental Decision Support System (WEDSS), which helps to prioritize local watersheds in terms of environmental sensitivity using multiple criteria identified by planners and local government staff in the city of Columbia, and Boone County, Missouri.

The WEDSS is an example of a way to run spatial models over the Web and represents a significant increase in capability over other WWW-based GIS applications that focus on database querying and map display. The WEDSS seeks to aid in the development of agreement regarding specific local areas deserving increased protection and the public policies to be pursued in minimizing the environmental impact of future development. The tool is also intended to assist ongoing public information and education efforts concerning watershed management and water quality issues for the City of Columbia, Missouri and adjacent developing areas within Boone County, Missouri.

Development of an Integrated Range Management Decision Support System

Integrated range management decision support system (IRMDSS) has been developed by Sugumaran (2002) to provide an interactive decision support tool for forest planners in the Western Ghats of India, which is one of the ecologically important areas of the Earth. The IRMDSS interface combines individual technologies such as remote sensing, a geographic information system (GIS) and a knowledge based system. The prototype IRMDSS was developed by customizing ARCVIEW GIS software and coupling it other software using dynamic link libraries. The IRMDSS prototype provides forest planners and managers a better means of organizing, accessing, and evaluating a wide range of information and alternative strategies for effective forest planning and management in the study area.

Urban Growth Modelling on the Web: A Decision Support Tool for Community Planners

Compasa and Sugumaran (2004) have develop a simple web-based urban growth model and visualization tool for St. Charles County's planning and zoning department to use in urban growth planning and management. They have demonstrated a web-based decision support tool for modelling urban growth under a variety of user-defined conditions utilizing multi-temporal (1984, 1992 and 2000), multi-seasonal (leaf off and leaf on), and multi-spectral Landsat images, along with several other data layers such as transportation networks, demography, topography, and zoning.

Being web-based and operational, this model is accessible not only to urban planners but also to non-technical users from any Internet browser in contrast to other similar tools available locally on personal computers. The paper concludes by discussing the model's current limitations and future directions. The model is located at http://maps.cares.missouri.edu/analysis/urban.

19

SRPIS

An Overview

SRPIS, Sathanur Reservoir Project Information System, enables the user to store, analyze, and display spatial information on the Sathanur Reservoir Project (SRP) as maps and attribute data.

The database module of *SRPIS, SRPIS-DBM*, enables the user to manipulate and update the database of SRP on a real time basis. SRPIS has 31 pre-designed forms which enable cognition of a wide variety of data. The user can browse the existing information through *forms, tables, and reports*, as one does when using MS Access. Also, the user can print SRPIS information as *reports, tables*, and *graphs*. In other words, even as the user works in a GIS environment, s/he has all the dynamism, flexibility, fluency, and user-friendliness that distinguishes MS Access.

Another unique feature of SRPIS is a novel EIA sub-module which is integrated with the software's database module. The environmental Impact assessment ratings of the Sathanur reservoir project can be directly fed into this module by various observers: experts, end users, and other groups. The personal details of each observer are stored in the table *EIA_Personaldetails*, and the EIA ratings are stored in a yet another table called *EIA-Matrix*. In the latter, the values are saved cumulatively for each village and for each category of the observer.

The map module of SRPIS, *SRPIS-Mapper*, has been designed specially for GIS requirements. The user can browse through various facets of SRPIS as map layers, thematic features, the attribute data related to these maps, hydraulic designs of Sathanur reservoir, and information on tourist interests in-and-around Sathanur. All the maps have been designed to be self explanatory, with emphasis on clear legends and map scales. However, in SRPIS one can't edit, manipulate, or create new thematic maps. To get around this shortcoming, we have incorporated options for installation of *MapInfo Proviewer 6.5* and downloading of *SRPIS Map Data* from the SRPIS installation CD-ROM disk.

MapInfo Proviewer 6.5, a GIS application, enables opening, displaying, printing, and limited manipulation of *Map Info tables ('.tab')* and workspaces that are provided with the SRPIS CD-ROM disk as *SRPIS Map Data*. Moreover, if the user has *MapInfo 5.5* or later versions installed, s/he can have the full power of editing, manipulating, and creating his/her own thematic maps using the *SRPIS Map Data*.

SRPIS has been designed to work in Windows OS and is compatible with Windows 95/98/2000/ME/NT/XP. It requires easy-to-find hardware comprising of Pentium III processor, 32 MB RAM and 1 GB free hard disk space. With Internet Explorer 5.0 (or higher version), greater benefit can be derived from SRPIS-Mapper.

Some of the unique features of SRPIS include:

— a novel EIA sub-module, which displays the EIA ratings by various experts and user groups.

— User-friendly interface with *on-the-fly* options for
  - viewing data as *tables, forms* and *reports* in MS Access
  - viewing SRPIS map data using Internet explorer

— dynamic viewing of maps with layer control facility, and hyperlinking of maps and attribute data

— zoom and pan features for better viewing of the maps.

20

Architecture of SRPIS (User's Guide)

SRPIS consists of four major modules: Installation Module *(SRPIS-ISM)*, Database Module *(SRPIS-DBM)*, Mapping Module *(SRPIS-Mapper)*, and SRPIS Accessories (SRPIS-ACC), consisting of *SRPIS Map Data* and *MapInfo ProViewer 6.5*.

Installation Module: ***SRPIS-ISM***

This module enables the user to install the various components of SRPIS: Installation Module *(SRPIS-ISM)*, Database Module *(SRPIS-DBM)*, Mapping Module *(SRPIS-Mapper)*, and SRPIS Accessories (SRPIS-ACC), consisting of *SRPIS Map* Data and *ProViewer 6.5* (Figure—20.1 and 20.2).

Installation Procedure

To install SRPIS and its components follow these steps:

1. *Start menu* > *Run* > type or browse: G:\SRPIS_INSTALLATION\SetUp.exe 'G' being the location of the CD-ROM drive, if not substitute 'G' with the current location of the CD-ROM drive where the SRPIS CD is placed.
2. Click on the *SRPIS* button to install complete programme of *SRPIS*.
3. Click on the *Install Map Data* button to copy the files of SRP maps.

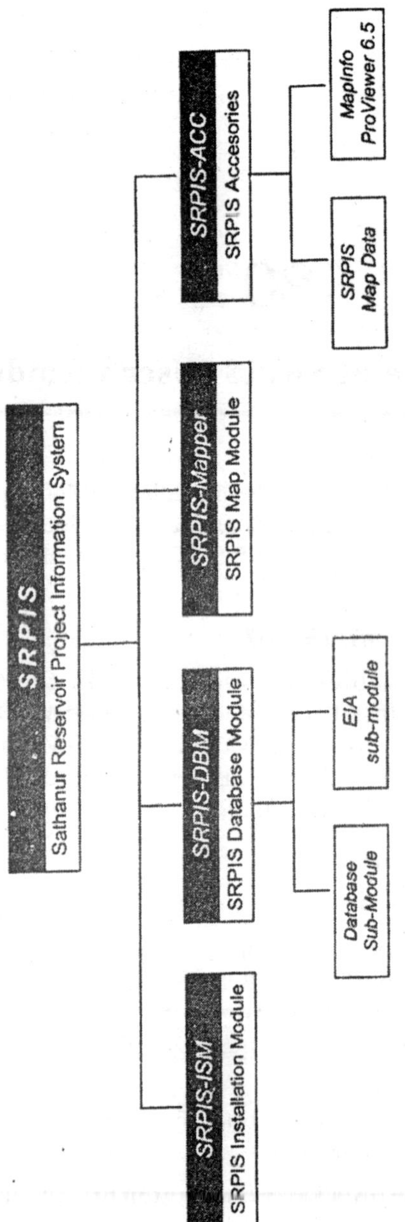

Fig. 20.1: Major modules of SRPIS

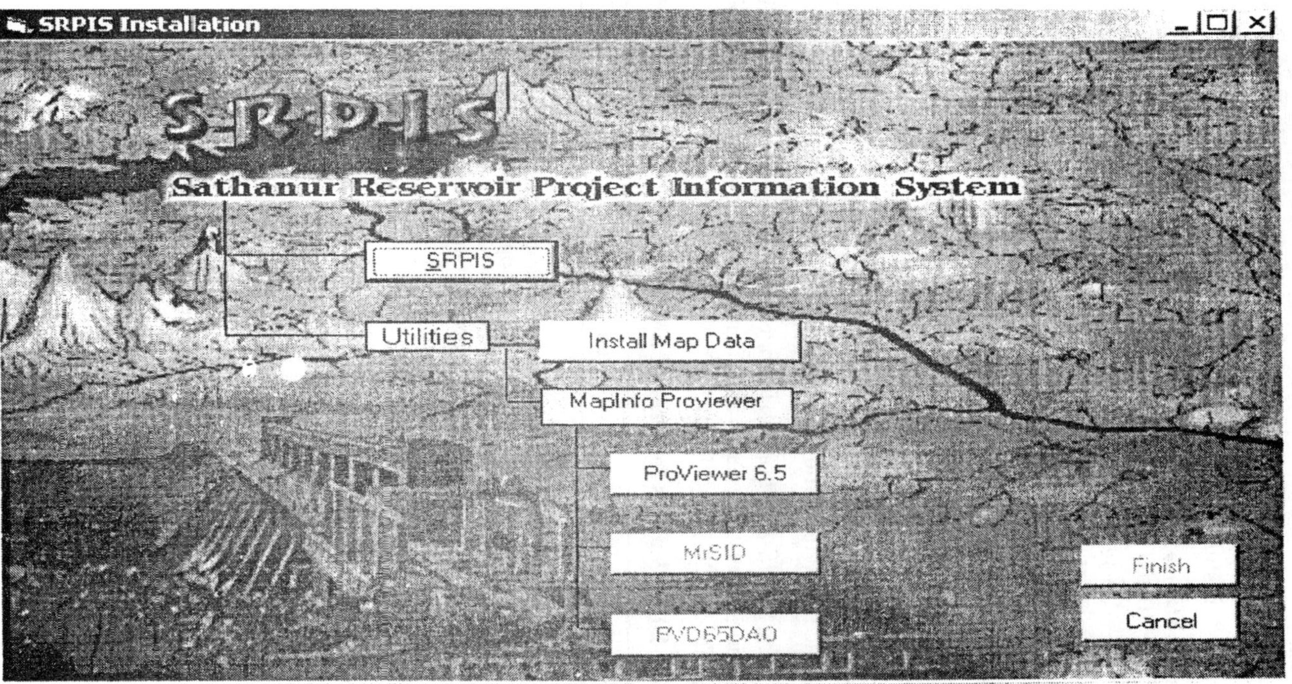

Fig. 20.2: The GUI of installation module of SRPIS, *SRPIS-ISM*

4. To install *ProViewer* click on the *ProViewer* button. Once *ProViewer* has been installed proceed with the installation of *MrSID* and *PVD65DA*.

Note: For *MapInfo ProViewer* to function properly the rest of its components—Mr SID and PVD65DAO have to be installed.

5. To terminate the process of installation click Cancel button.

6. Click Finish after completing installation.

Database Module: *SRPIS-DBM*

This module enables the users to store, update, edit, and retrieve the SRPIS data as-and-when required. The module also provides dynamic link with MS Access. The user can perform SQL queries and use all the built-in functions of MS Access using this dynamic link within the SRPIS environment.

The *SRPIS-DBM* opens with a Graphic User Interface (GUI) consisting of specific menus and icons, which have been described in the Table—20.1, and Figure—20.3a.

Table—20.1 Various features of SRPIS-DBM

	Menu	Icon	Functionality
File			
	Open		The menu command opens the list *of forms* The icon on clicking opens a dialog box in which the name of the form has to be typed
	Save		Saves the contents of the current *form*
	Close	×	Closes the active *form*
	Exit	×	Exit from SRPIS
View			
Database			
	Table		Opens the MS Access table of the active *form*
	Report		Opens the MS Access report of the active *form*
	Maps		Opens the SRPIS-*Mapper* module
Windows			
Tile			
	Horizontal	×	Tiles all the *forms* horizontally
	Vertical	×	Tiles all the *forms* vertically
Cascade		×	Arranges the open *forms* in the cascading fashion
Help			
	Contents	×	Opens the contents help files to browse
	Credits	×	Opens the credits frame of SRPIS

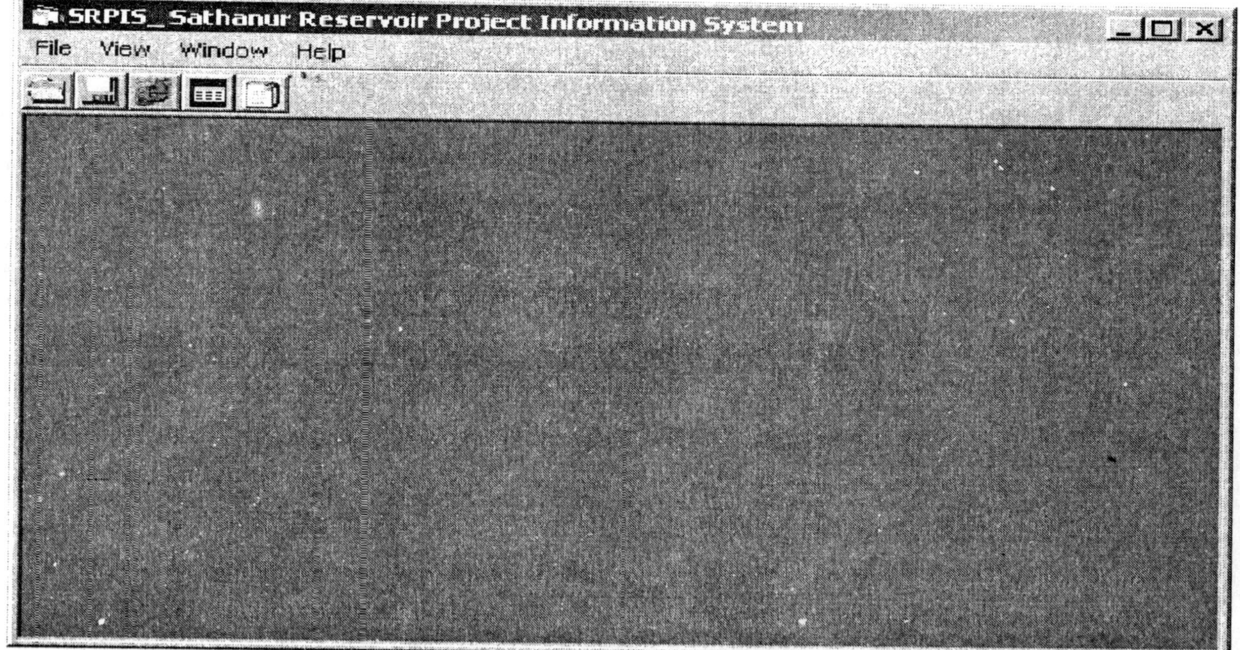

Fig. 20.3 (a): The graphic user interface (GUI) of the SRPIS database module (*SRPIS-DBM*)

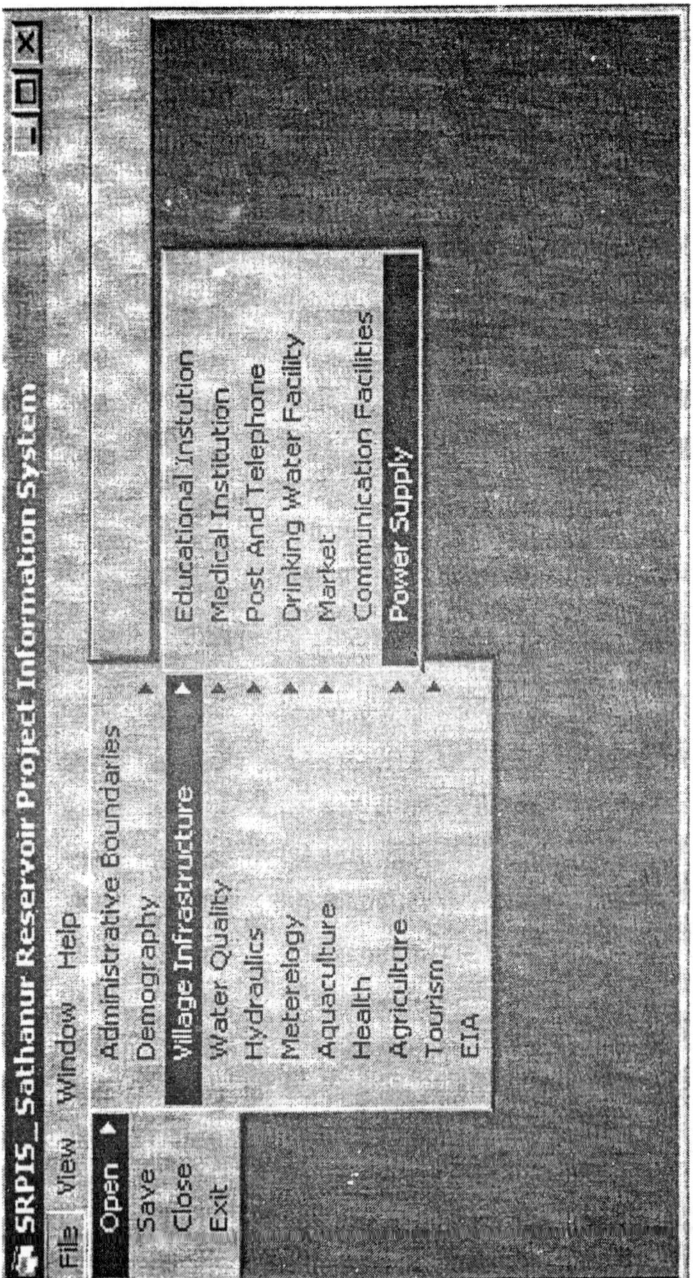

Fig. 20.3(b): The cascading *open* dialog box of *SRPIS*

SRPIS-DBM consists of 31 pre-designed forms for the manual entry of the data pertaining to various aspects of SRPIS. The table structure of these forms have been described in the chapter 8, (Volume 1) *Reference Guide.*

An overview of the various features of *SRPIS-DBM* and their functionalities have been presented as Figure—20.4. The behaviour of all these features is identical to that of a standard Windows program except that of open menu and *open* icon.

Opening the forms. The *open* menu displays a list of various forms available in a cascading fashion (Figure—20.3b). The user has to choose one of these forms to view it. The *open* icon would display a dialog box, in which the name of the *form* has to be entered for opening it.

All the forms are designed with appropriate validation checks, suggestion dialog boxes and warning dialog boxes. The behaviour and function of these forms is similar to the standard MS Access forms. The latter being time tested and established as user friendly.

Information retrieval. The information stored in the database can be retrieved and displayed as *forms, reports,* and *tables* (Figure—20.5).

Viewing data in MS Access. The information stored in the SRPIS database can be viewed by selecting: *View* > *Database* > *Table* or *View* > *Database* > *Report* (Figure—20.6). The user can access all the features of MS Access such as sorting of data, viewing the data as graphs, saving the table as a new file, renaming the tables, and copying the contents of the table to another table or a new database.

Note: Please don't rename the original tables of SRPIS database. Once the tables are renamed, SRPIS can't access these new tables. It is advised to save the renamed tables in a new database file.

Viewing SRPIS Maps. The interface of the database module provides link for opening the maps of SRPIS using Internet Explorer (IE). To open the SRPIS maps select: *View* > *Maps* > *Run SRPIS Mapper* or click on the icon of IE.

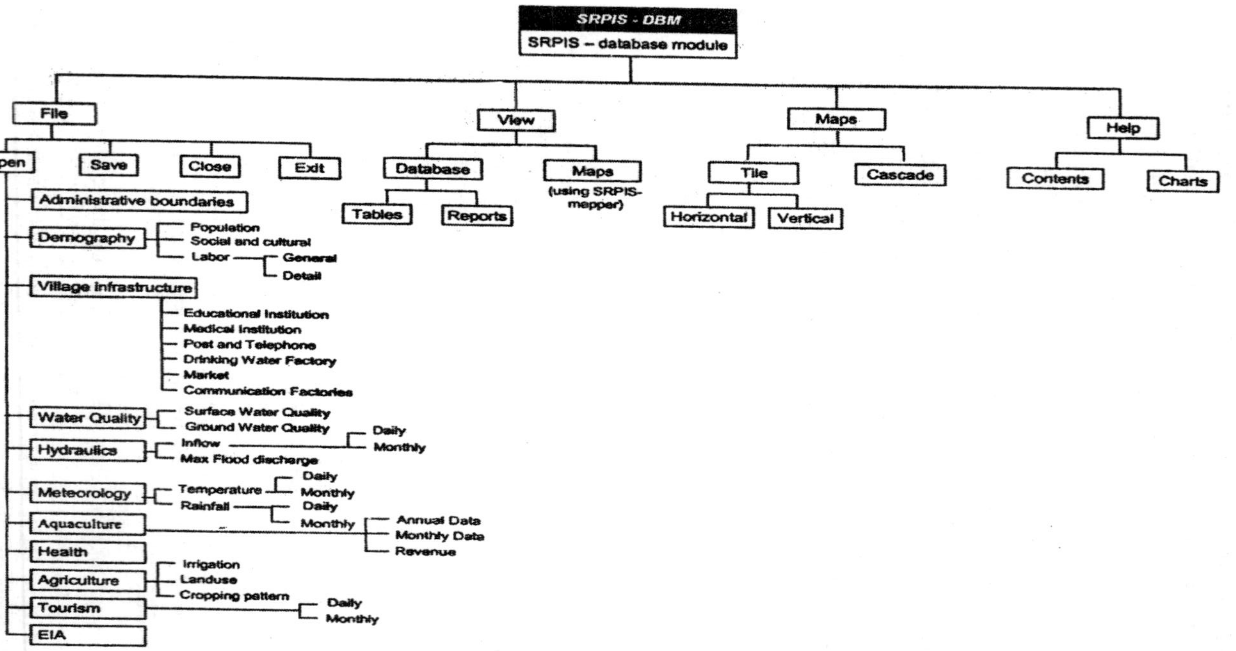

Fig. 20.4: An overview of the various features of SRPIS database module, *SRPIS-DBM*

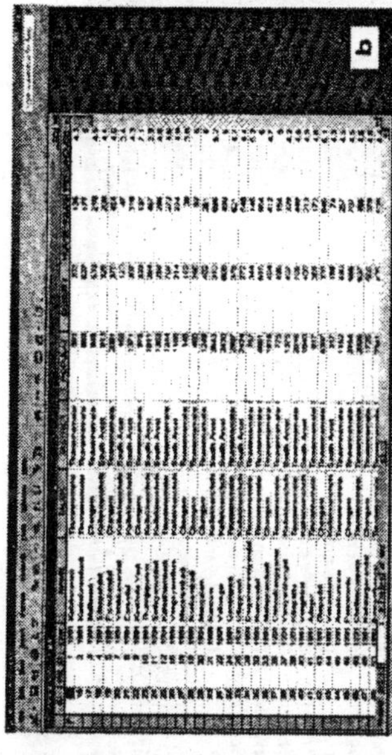

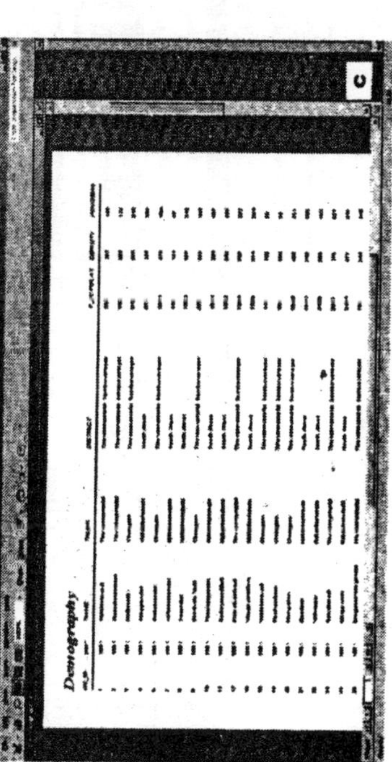

Fig. 20.5: Information retrieval using *SRPIS-DBM* as (a) *forms*, (b) *tables*, and (c) *reports*

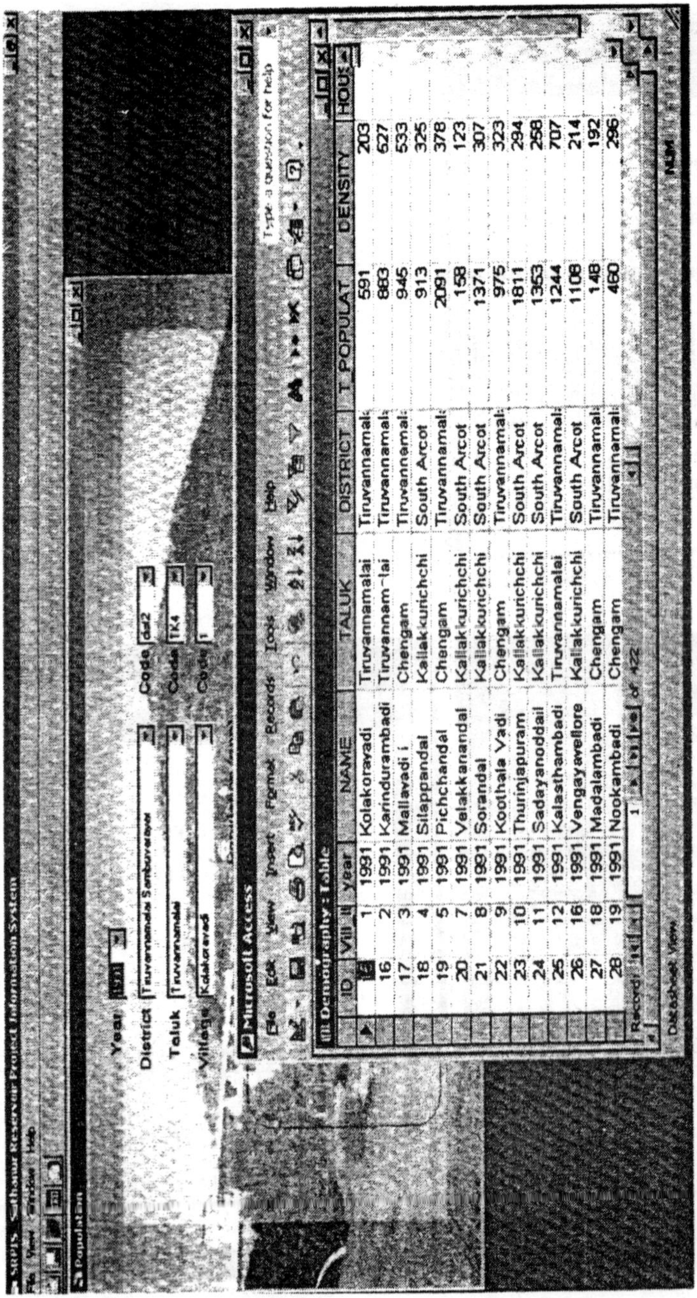

SRPIS Sathanur Reservoir Project Information System

Population

Year: 1991

District: Tiruvannamalai Sambuvarayar — Code: dst2

Taluk: Tiruvannamalai — Code: TK4

Village: Kolakoravadi — Code: 1

Microsoft Access

Demography : Table

ID	Vill_I	year	NAME	TALUK	DISTRICT	T_POPULAT	DENSITY	HOUS
[illegible]	1	1991	Kolakoravadi	Tiruvannamalai	Tiruvannamal:	591	203	
16	2	1991	Karindurambadi	Tiruvannam~lai	Tiruvannamal:	883	627	
17	3	1991	Mallavadi i	Chengam	Tiruvannamal:	945	533	
18	4	1991	Silappandal	Kallakkurichchi	South Arcot	913	325	
19	5	1991	Pichchandal	Chengam	Tiruvannamal:	2091	378	
20	7	1991	Velakkanandal	Kallakkurichchi	South Arcot	158	123	
21	8	1991	Sorandal	Kallakkurichchi	South Arcot	1371	307	
22	9	1991	Koothala Vadi	Chengam	Tiruvannamal:	975	323	
23	10	1991	Thurinjapuram	Kallakkurichchi	South Arcot	1811	294	
24	11	1991	Sadayanoddail	Kallakkurichchi	South Arcot	1353	258	
25	12	1991	Kalasthambadi	Tiruvannamalai	Tiruvannamal:	1244	707	
26	16	1991	Vengayavellore	Kallakkurichchi	South Arcot	1108	214	
27	18	1991	Madalambadi	Chengam	Tiruvannamal:	148	192	
28	19	1991	Nookambadi	Chengam	Tiruvannamal:	460	296	

Record: 1 of 422

Datasheet View

Fig. 20.6: The SRPIS database module (*SRPIS-DBM*) providing dynamic link with MS Access

Window options. This option, which is available in the menu, helps one to arrange the forms that are open. These forms can be tiled (horizontally or vertically) or arranged in a cascading fashion. This feature is similar to the options that are available in most of the standard Windows-based software.

Help files. SRPIS has well-built help files featuring the contents and credits of the software development. The help files are illustrated wherever necessary and the user-commands are provided legibly.

EIA Sub-module

This sub-module of SRPIS is a novel attempt to automate the process of environmental impact assessment (EIA) of Sathanur Reservoir Project (SRP). This sub-module has two tables: *EIA_matrix* and the *EIA_personal* details.

The EIA form, listed in the *open* menu of the database module (SRPIS-DBM), opens with a window featuring various fields related to the details of location of the village, taluk, and district; the personal details of the evaluator—whether end-user/expert/other; name, age, occupation, education, income, and land holding (Figure—20.7). The user can clear these fields and re-enter freshly or can proceed to the next page by clicking on the next button.

The subsequent page contains the questionnaire for various indicators: agriculture, irrigation, maintenance of canals and tanks, and socio-economics (Figure—20.8). The status of these indicators can be selected by clicking on the *radio* buttons provided against various classes: worse (-3), very bad (-2), bad (-1), no difference (0), Good (1), very good (2), excellent (3). The *radio* button ensures that only one of these classes is selected.

On completing the questionnaire, the values of the selected classes are added cumulatively for each indicator, for each village, and the category of the evaluator in the *EIA_Matrix* table (Figure—20.9). The final report of the *EIA_Matrix* would always display the current value - the cumulative value of the indicators for each village. The personal details of the evaluators is stored in a separate table titled *EIA_Personal* (Figure—20.10).

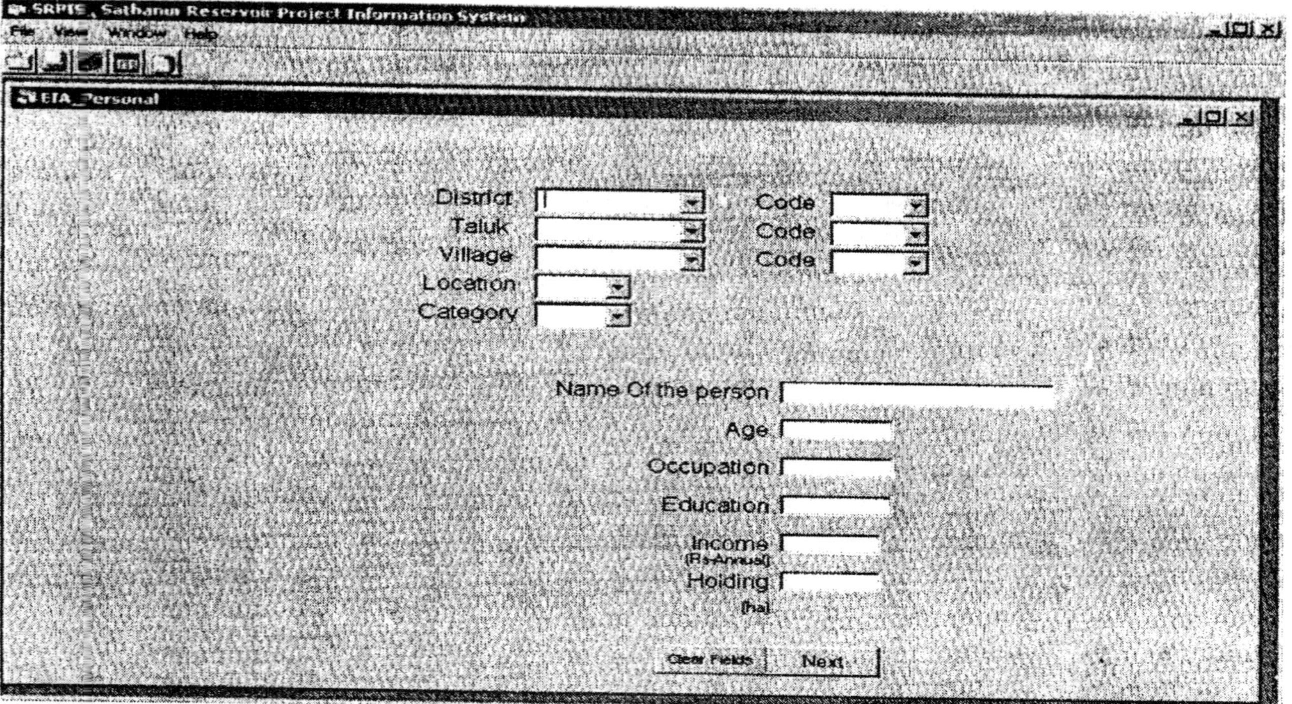

Fig. 20.7: The *form* of EIA sub-module with provision for cataloguing the personal details of the evaluator

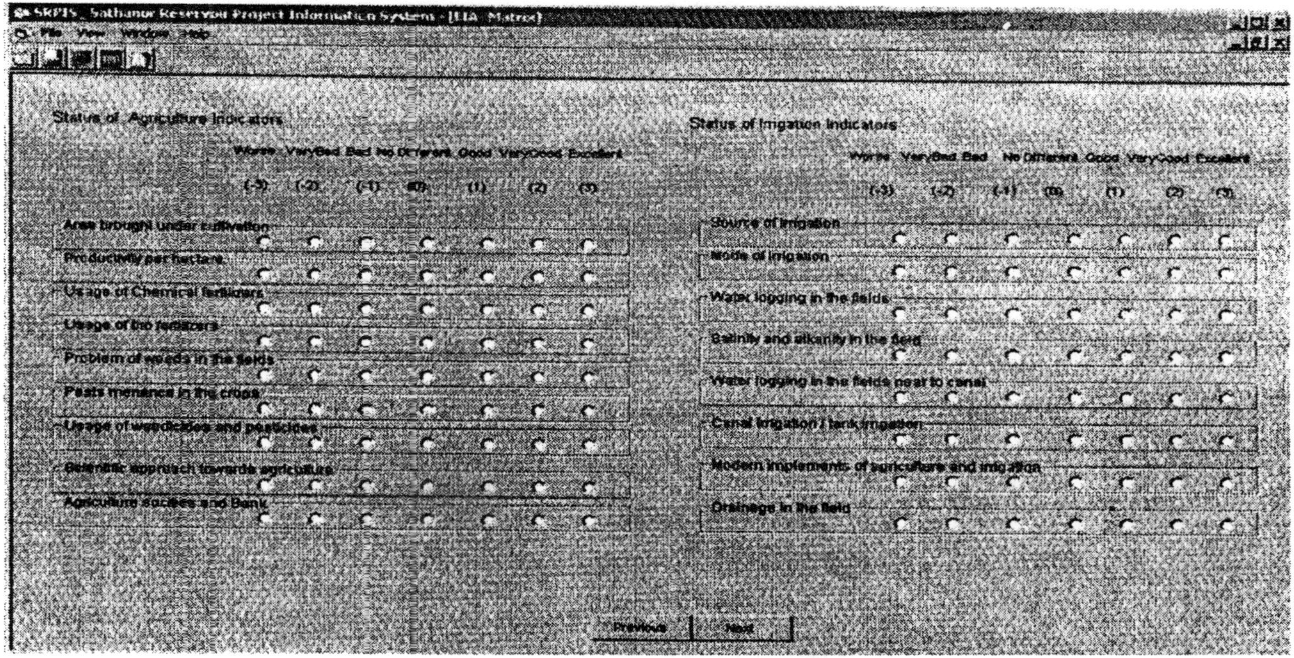

Fig. 20.8: The *form* of EIA sub-module with the objective questionnaire for evaluating the various socio-environmental indicators of SRP

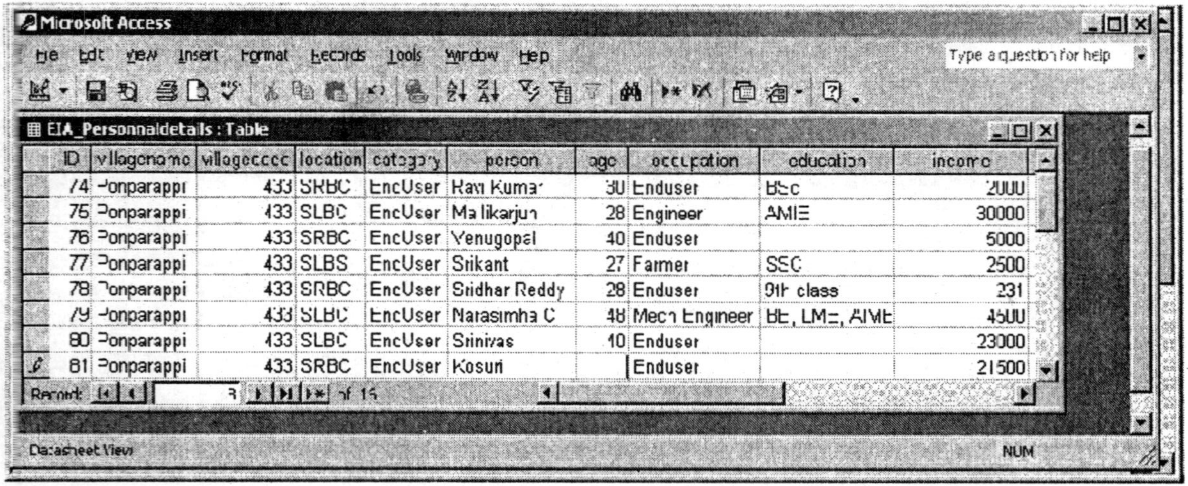

ID	villagename	villagecode	location	category	person	age	occupation	education	income
74	Ponparappi	433	SRBC	EndUser	Ravi Kumar	30	Enduser	BSc	2000
75	Ponparappi	433	SLBC	EndUser	Mallikarjun	28	Engineer	AMIE	30000
76	Ponparappi	433	SRBC	EndUser	Venugopal	40	Enduser		5000
77	Ponparappi	433	SLBS	EndUser	Srikant	27	Farmer	SSC	2500
78	Ponparappi	433	SRBC	EndUser	Sridhar Reddy	28	Enduser	9th class	231
79	Ponparappi	433	SLBC	EndUser	Narasimha C	48	Mech Engineer	BE, LME, AIME	4500
80	Ponparappi	433	SLBC	EndUser	Srinivas	10	Enduser		23000
81	Ponparappi	433	SRBC	EndUser	Kosuri		Enduser		21500

Fig. 20.9: The *table* of EIA sub-module, *EIA personal*

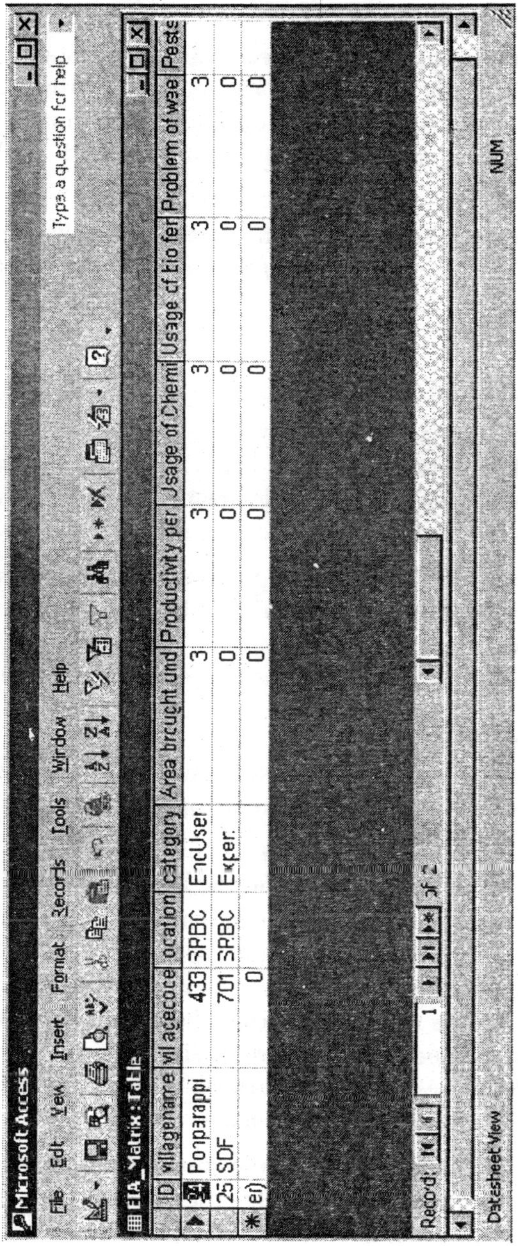

ID	villagename	vil agecode	ocation	category	Area brcught und	Productivity per	Jsage of Chemi	Usage cf bio fer	Problem of wae	Pests
24	Ponparappi	433	SPBC	EndUser	3	3	3	3	3	
25	SDF	701	SPBC	Exper.	0	0	0	0	0	
er)		0			0	0	0	0	0	

Fig. 2C.10: The *table* of EiA sub-module, *EIA_Matrix*

The Mapping Module: *SRPIS-Mapper*

This module has been designed to browse through various facets of SRPIS: map layers, thematic maps, the attribute data connected to these maps, hydraulic designs of Sathanur reservoir, and information on tourist interests around Sathanur (Figure—20.11).

SRPIS-Mapper has six sub-units (hyper-linked html pages): *Map layers, Thematic maps, More info, About us, Help, and Credits* (Figure—20.12).

Map Layers

This sub-unit of SRPIS-Mapper has been designed to display the political and land-use/land-cover maps of Ponnaiyar basin, and the catchment and command areas of Sathanur reservoir. These maps have been presented as interactive and individual map layers, which can be displayed in combinations.

The browser of *Map layers* is composed of four control units and three display frames: *main map control unit, index map control unit, SRP map control unit, scale display unit, index map display frame, main map display frame, and legend display frame* (Figure—20.13).

Main Map Control Unit

This unit has zoom-in, zoom-out, and pan control buttons. These features can be implemented only on the maps rendered in the *main map display frame*.

Zoom-in /zoom-out. Clicking on these buttons changes the cursor into cross-hair shape. The map, rendered in the *main map display frame,* can be zoomed-out (reducing the map size) or zoom-in (enlargement in the map size) by clicking the left-hand button of the mouse on the map. Scale factor, the number of times the map has been zoomed-in or zoomed-out, is displayed in the *scale display frame*. Also, scale bar and the north direction are indicated in the right bottom corner of all the maps.

Pan. On selecting this button the cursor shape changes to a *pointing hand*. To move the map across the browser, click the left-hand button of the mouse, hold the button, and move the mouse cursor to the desired position.

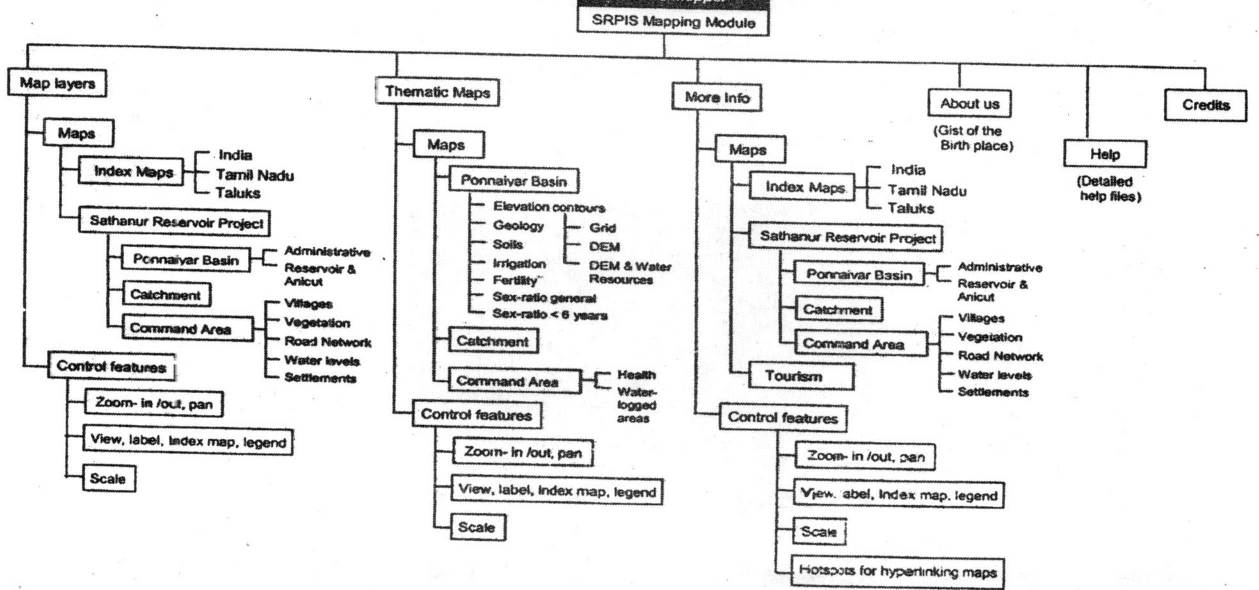

Fig. 20.11: An overview of the mapping module of SRPIS, *SRPIS-Mapper*

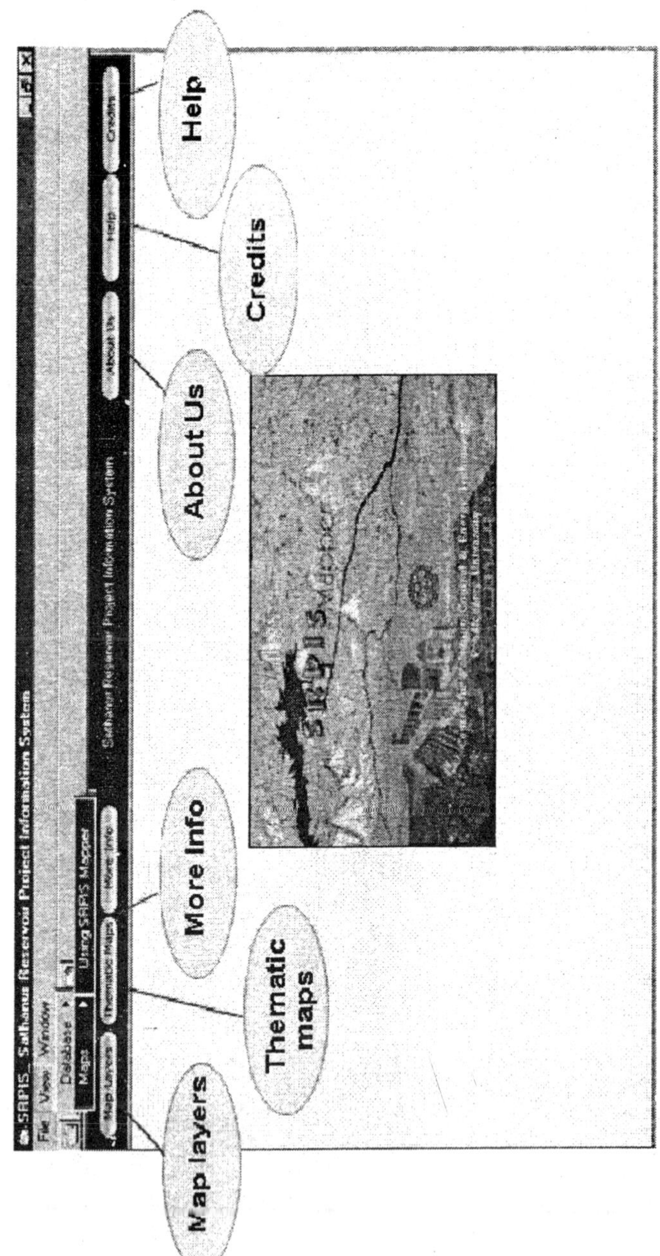

Fig. 20.12: The *SRPIS-Mapper* module. Please Note the sub-titles of *SRPIS-Mapper*

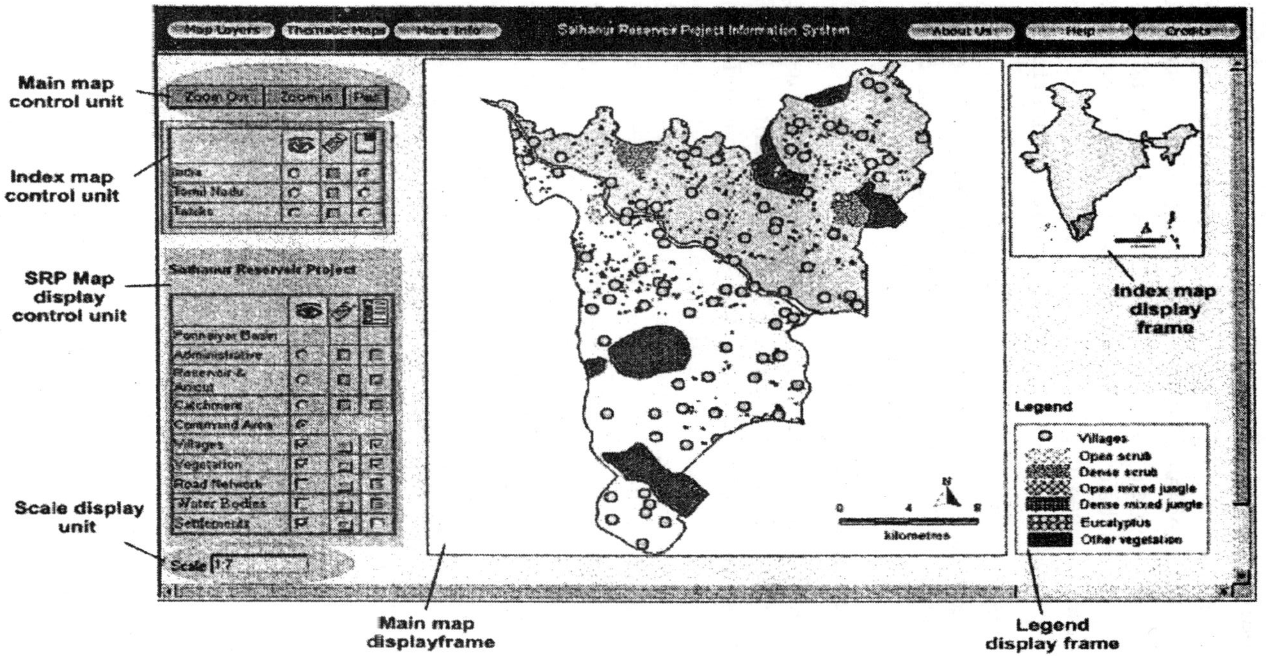

Fig. 20.13: The control units and display frames of *SRPIS-Mapper module*

Index Map Control Unit

This unit has the index maps—political maps of India, Tamil Nadu, and taluks of Tamil Nadu—listed on the left-hand side. To view these maps in the main map display frame select the radio button ⊙ provided below the view icon. Only one of the index maps can be displayed in the *main map display frame*. The map labels can be displayed on the map by clicking in the *check box* below the lable icon.

To place the maps in the Index map frame select the *radio button* provided below the *index* icon. Only one map can be displayed in the *index map display frame* at a time.

SRP Map Control Unit

Due to the large variance in the *scale* of maps, for instance, the maps of the scale of a river basin (Ponnaiyar basin) and that of the village maps (of command area), it is not possible to display both these maps with equal resolution at once. Hence, the maps of SRP have been grouped in three sets: *Ponnaiyar Basin, (SRP) catchment*, and the *(SRP) command area*. The group *Ponnaiyar basin* consists of an administrative map and a reservoir-anicut map. The catchment group consists of a map of the Sathanur reservoir catchment. The other group *command area* consists of maps featuring *villages, vegetation, road network, water bodies* and *settlements*. Only one map among these groups can be displayed at a time in the main map display frame. To select a group of maps to be viewed, click on the *radio button* ⊙ provided below the view icon.

The maps under the *command area* can be selected in combinations of *'n'*, which would be reflected in the *main map display frame*.

The labels can be displayed by clicking in the *check boxes* against the *label icon* for *Ponnaiyar basin and catchment*. Whereas the labels for the *command area* maps can be displayed by clicking the *push button*, provided below the label icon. Once the button has been clicked, the labels will be displayed till the cursor/mouse lingers on the button. This is one of the unique features of SRPIS.

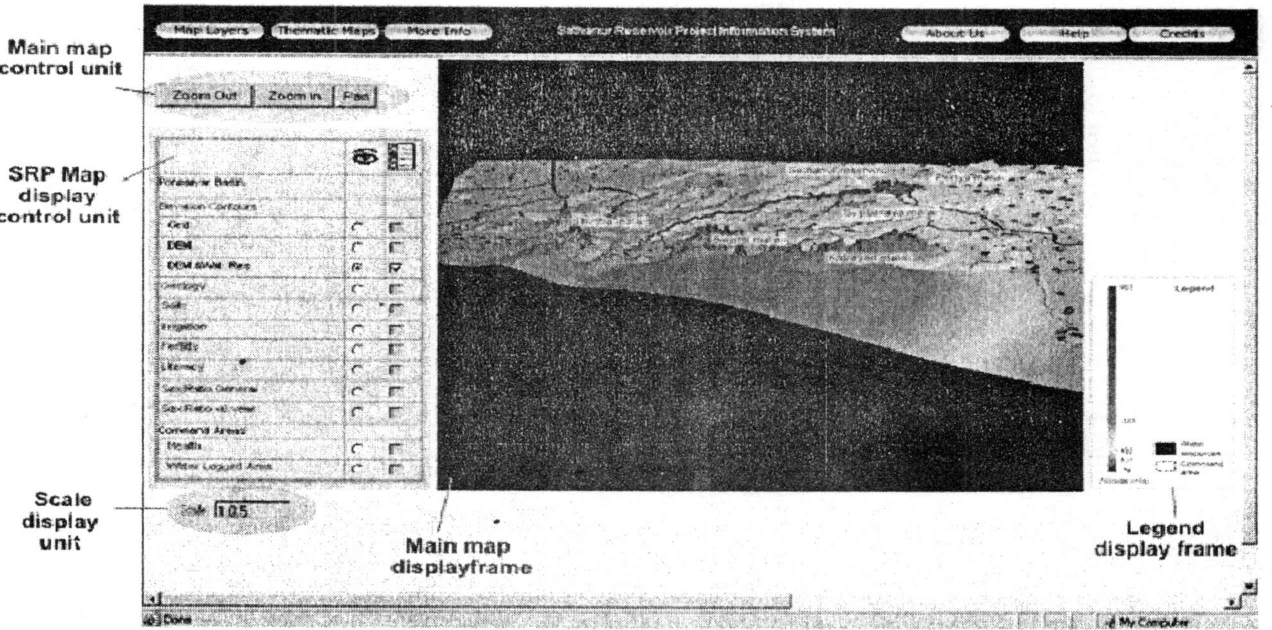

Fig. 20.14: The control units and display frames of the *thematic maps browser*

Legends. Legends for the maps under *SRP map control* unit can be displayed by selecting the check boxes below the *legend icon*, against the active map.

Thematic Maps

This sub-unit of *SRPIS-Mapper*, has been designed to display the thematic maps of Ponnaiyar basin and the command areas of Sathanur reservoir project.

The browser of *thematic maps* is composed of two control units and three display frames: *main map control unit, thematic map control unit, scale display frame, main map display frame,* and *legend display frame* (Figure—20.14).

Main Map Control Unit

This unit has zoom-in, zoom-out, and pan control buttons. These features can be implemented only on the maps rendered in the *Main map display frame.*

Zoom-in/zoom-out. Clicking on these buttons changes the cursor into *cross-hair shape*. The map, rendered in the *main map display frame,* can be zoomed-out (reducing the map size) or zoom-in (enlargement in the map size) by clicking the left-hand button of the mouse on the map. Scale factor, the number of times the map has been zoomed-in or zoomed-out, is displayed in the *scale display frame*. Also, scale bar and the north direction are indicated in the right bottom corner of all the maps.

Pan. On selecting this button the cursor shape changes to a *pointing hand*. To move the map across the browser, click the left-hand button of the mouse, hold the button, and move the mouse to the desired position.

Thematic Map Control Unit

This unit has two sub-groups of maps designed for Ponnaiyar basin and the SRP command area. The sub-group *Ponnaiyar basin* consists of elevation contours (as grid, digital elevation model (DEM), and water resources), geology, soils, irrigation, literacy, sex-ratio general, and sex-ratio below 6 years. The other sub-group of maps consists of health and water-logged areas of *command area*.

To display a thematic map, click on the radio button ⊙ provided below the view icon. Only one map can be displayed in the main *map display frame at a time.*

Legends. Legends for each sub-group can be displayed by selecting the check boxes below the legend icon, against the active map.

More Info

The browser of *more info* is composed of three control units and four display frames: *index map control unit, SRP map control unit, Scale display unit, Main map display frame, Index map display frame, Other info hyper-link frame,* and *legend display frame* (Figure—20.15).

Index Map Control Unit

This unit has the index maps—political maps of India, Tamil Nadu, taluks of Tamil Nadu—listed on the left-hand side. To view these maps select the *radio* button ⊙ provided below the *view* icon. Only one of the index maps can be displayed in the *main map display*. The map labels can be displayed on the map by clicking in the *check box*.

To place the maps in the *index map frame* select the *radio* button provided below the *index icon*. Only one map can be displayed in the *index map frame* at once.

Note: The hyperlinks work only when the *labels*, and the *legends* are turned off.

SRP Map Control Unit

Similar to the *Map layers* browser, the maps of this unit have been grouped into three categories: *Ponnaiyar Basin, (SRP) catchment*, and the *(SRP) command area.*

The *Ponnaiyar basin* group consists of an administrative map, and a reservoir-anicut map. The administrative map has attribute information of the demographic details of the various states, across which the Ponnaiyar basin spreads. This attribute information can be accessed by clicking in the map regions with *hotspots*. The shape of the cursor turns into a *pointing hand* when the mouse encounters

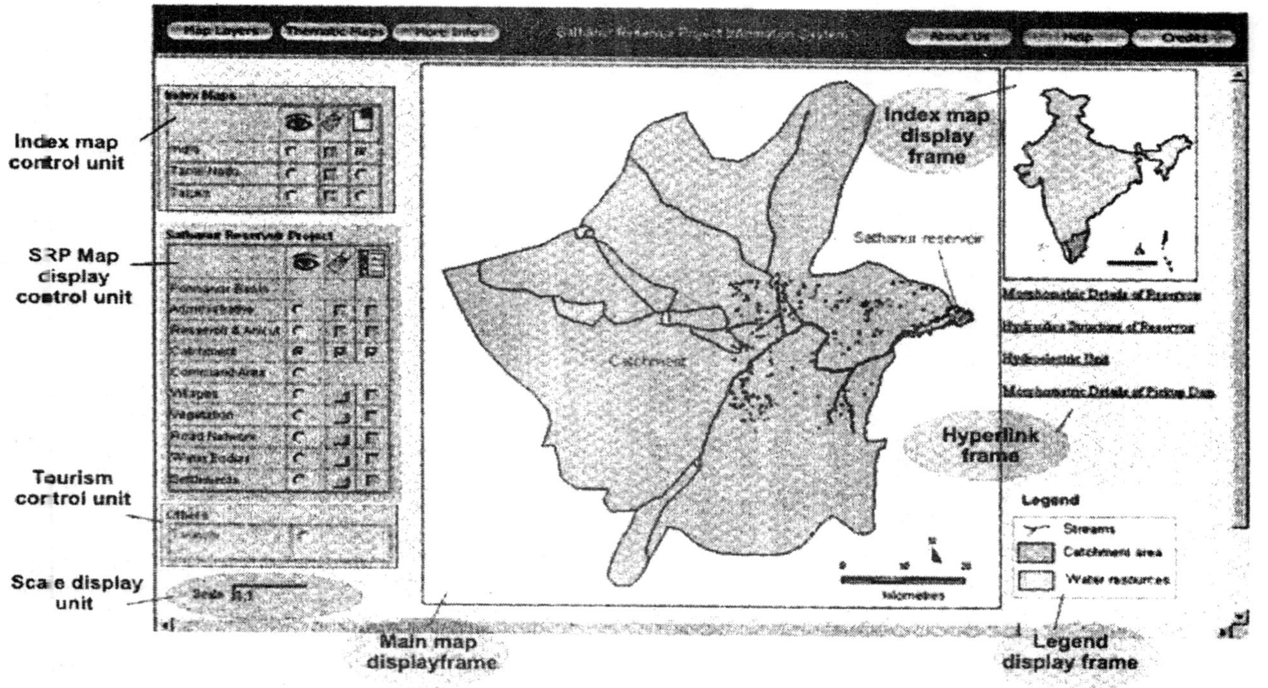

Fig. 3.15: The control units and the display frames of *More info* browser

a *hotspot*. These *hotspots* are hyperlinked with a separate web page which displays the attribute/additional information.

The reservoir and Anicut map contains attribute information pertaining to: (i) the catchment area of Ponnaiyar basin across the various states (ii) village-wise proposed area under canal irrigation in SLBC, and (iii) village-wise proposed area under canal irrigation in SRBC.

The *catchment* group displays the catchment map of Sathanur reservoir in the *main map display frame*. The hotspots of this map leads to the attribute information related to the catchment details of Ponnaiyar basin. Also, on the right hand side of the *main map display frame* a list of the other attribute information, which are hyperlinkable, is displayed: (i) morphometric details of Sathanur reservoir (ii) hydraulic structure (hydraulic designs) of the reservoir (iii) hydroelectric unit, and (iv) morphometric details of pickup dam. Clicking on these hyperlinks would open a new browser with further details.

The *command area* has maps as individual, interactive map layers: *villages, vegetation, road network, water bodies, and settlements*. The following attribute data has been attached with these individual map layers:

1. villages: (i) list of SRBC villages, district-wise (ii) list of SLBC villages, district-wise
2. vegetation: (i) types of vegetation and the area covered (ii) summary of the types of vegetation
3. road network: coverage of the road network
4. water resources: the number and the area covered by various water resources
5. settlements: spread of settlements

Note: The hyperlinks work only when the *labels*, and the *legends* are turned off.

Others: Tourism

This control unit displays a hyperlinked list of the places of tourist interest in-and-around Sathanur: *Sathanur, Tiruvannamalai,*

Yelagiri, Ginjee, and Others (Figures—20.16 and 20.17). The hyperlinked pages open in a new browser window. These html pages contain a brief introduction, site map, the places of tourist's interest with photographs, and some trivia. The gist of the contents of these html pages has been presented below:

Web page	Contents
Sathanur	Sathanur dam, Ponnaiyar reserve forest, greenery and park, crocodile farm, aquarium, swimming pool, mini-zoo
Tiruvannamalai	Sri Arunchaleshwar temple, Ramanashram, Sri Yogi Suratkumar ashram
Yelagiri	Yelagiri hills, Children's park, Velavan temple, Government herbal farm, Jalagamparai water falls, Shandy, fountain, mini zoo, government fruit farm, telescope house, Kavalur or Vainu pappu solar observatory, summer festival
Ginjee	Ginjee fort
Others	This is for updating any new site of tourist interest

Note: The hyperlinks work only when the *labels*, and the *legends* are turned off.

About us

The browser of About us contains a brief profile of *Centre for Pollution Control* and *Energy Technology*, where SRPIS has been designed and developed.

Help

The browser of *Help* is a comprehensive point-to-point user's guide for working with *SRPIS-Mapper* (Figure—20.18).

Credits

The browser of *Credits* has a list of the team involved in the making of SRPIS.

Accessories Module of SRPIS: *SRPIS-ACC*

This module consists of two components: *SRPIS Map Data* and *Map ProViewer 6.5.*

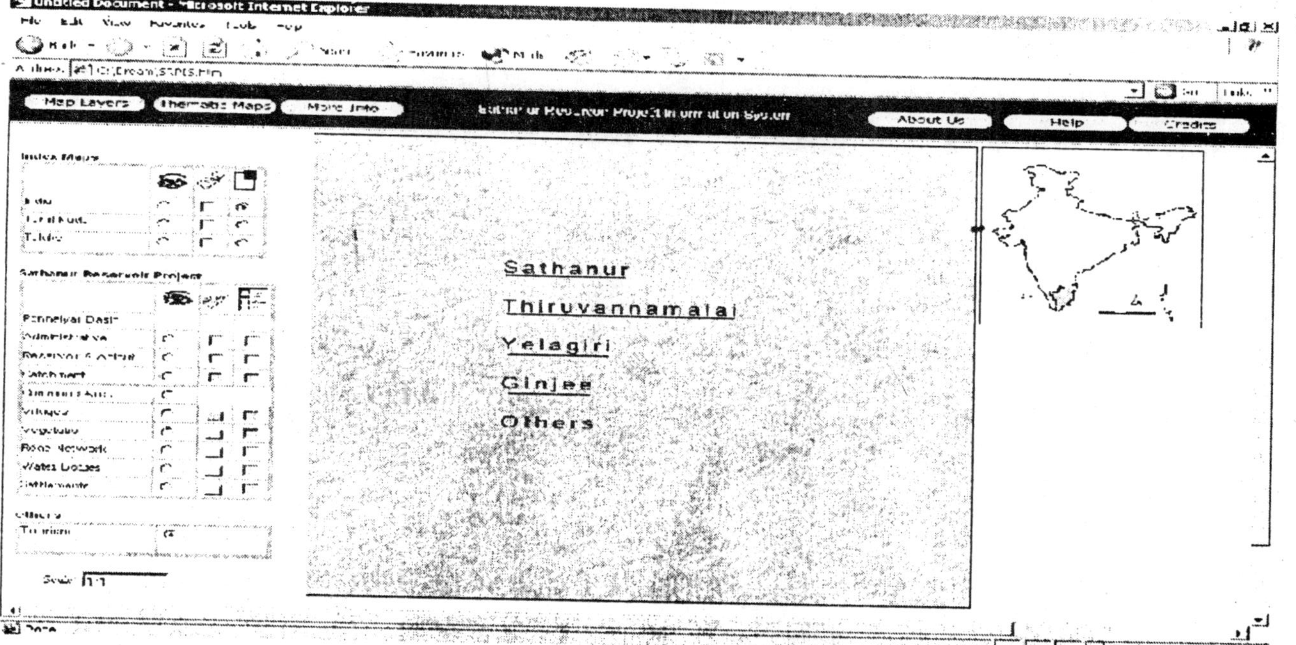

Fig. 20.16: The browser of *more info* displaying the *tourism* html page

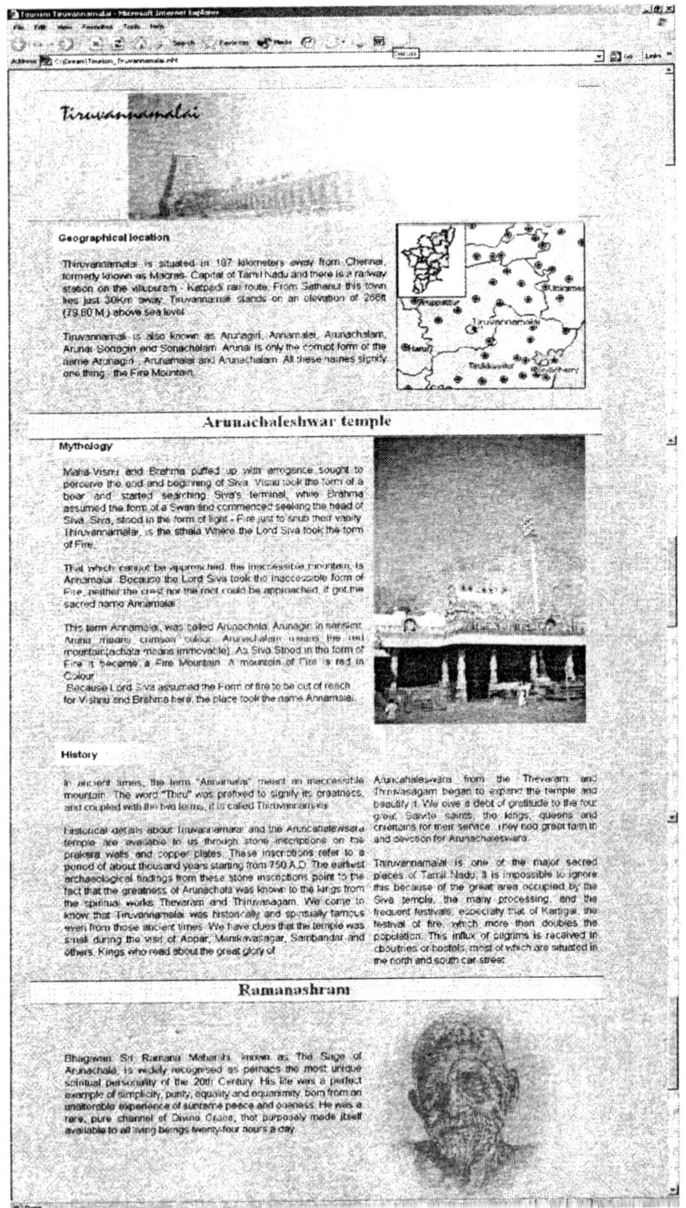

Tiruvannamalai

Geographical location

Thiruvannamalai is situated in 187 kilometers away from Chennai, formerly known as Madras- Capital of Tamil Nadu and there is a railway station on the villupuram - Katpadi rail route. From Sathanur this town lies just 30Km away. Tiruvannamalai stands on an elevation of 266ft (79.80 M.) above sea level.

Tiruvannamalai is also known as Arunagiri, Annamalai, Arunachalam, Arunai Sonagiri and Sonachalam. Arunai is only the corrupt form of the name Arunagiri , Arunamalai and Arunachalam. All these names signify one thing - the Fire Mountain.

Arunachaleshwar temple

Mythology

Maha-Visnu and Brahma puffed up with arrogence sought to perceive the end and beginning of Siva. Visnu took the form of a boar and started searching Siva's terminal, while Brahma assumed the form of a Swan and commenced seeking the head of Siva. Siva, stood in the form of light - Fire just to snub their vanity. Thiruvannamalai, is the sthala Where the Lord Siva took the form of Fire.

That which cannot be approached, the inaccessible mountain, is Annamalai. Because the Lord Siva took the inaccessible form of Fire, neither the crest nor the root could be approached, it got the sacred name Annamalai.

This term Annamalai, was called Arunachala, Arunagiri in sanskrit. Aruna means crimson colour. Arunachalam means the red mountain(achala means immovable). As Siva Stood in the form of Fire it became a Fire Mountain. A mountain of Fire is red in Colour.
Because Lord Siva assumed the Form of fire to be out of reach for Vishnu and Brahma here, the place took the name Annamalai.

History

In ancient times, the term "Annamalai" meant an inaccessible mountain. The word "Thiru" was prefixed to signify its greatness, and coupled with the two terms, it is called Thiruvannamalai.

Historical details about Tiruvannamalai and the Arunachalewsara temple are available to us through stone inscriptions on the prakara walls and copper plates. These inscriptions refer to a period of about thousand years starting from 750 A.D. The earliest archaeological findings from these stone inscriptions point to the fact that the greatness of Arunachala was known to the kings from the spiritual works Thevaram and Thiruvasagam. We come to know that Tiruvannamalai was historically and spiritually famous even from those ancient times. We have clues that the temple was small during the visit of Appar, Manikavasagar, Sambandar and others. Kings who read about the great glory of Aruncahaleswara from the Thevaram and Thiruvasagam began to expand the temple and beautify it. We owe a debt of gratitude to the four great Saivite saints, the kings, queens and chieftains for their service. They had great faith in and devotion for Arunachaleswara.

Thiruvannamalai is one of the major sacred places of Tamil Nadu. It is impossible to ignore this because of the great area occupied by the Siva temple, the many processing, and the frequent festivals, especially that of Kartigai, the festival of fire, which more then doubles the population. This influx of pilgrims is received in choultries or hostels, most of which are situated in the north and south car street.

Ramanashram

Bhagavan Sri Ramana Maharshi, known as The Sage of Arunachala, is widely recognised as perhaps the most unique spiritual personality of the 20th Century. His life was a perfect example of simplicity, purity, equality and equanimity born from an unalterable experience of supreme peace and oneness. He was a rare, pure channel of Divine Grace, that purposely made itself available to all living beings twenty-four hours a day.

Fig. 20.17: The hyperlink browser *Tiruvannamalai* of *tourism*

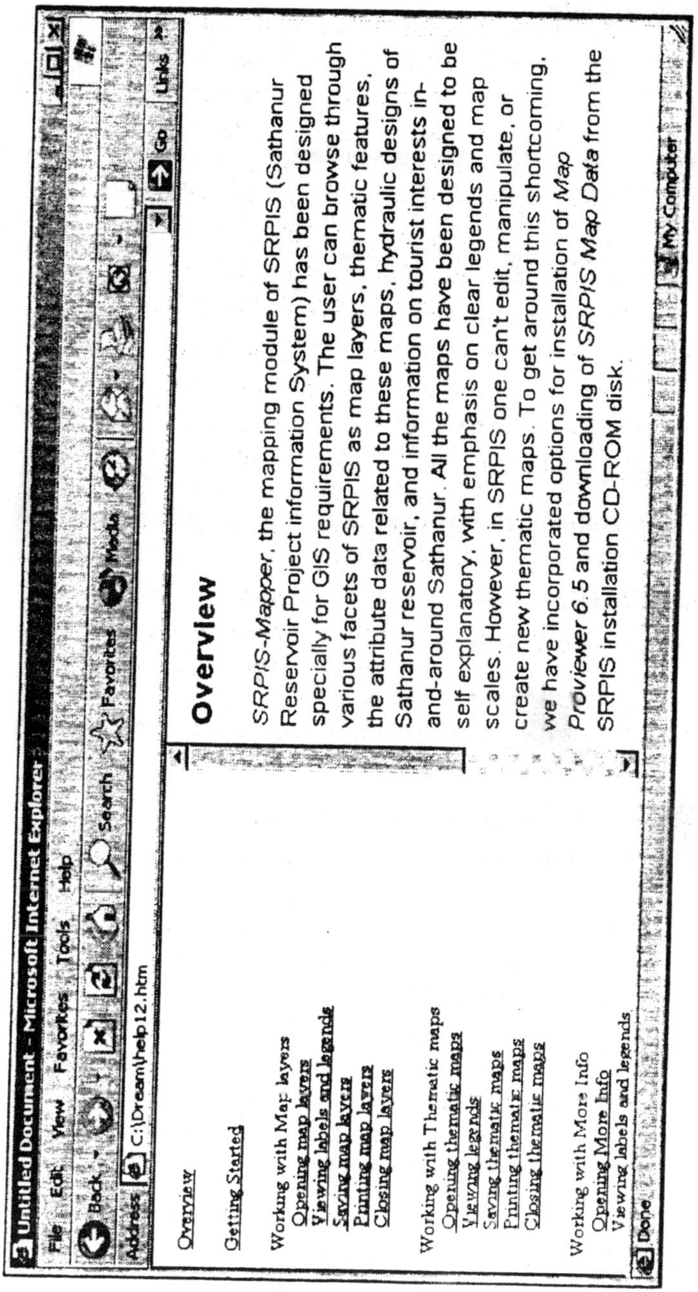

Untitled Document - Microsoft Internet Explorer

File Edit View Favorites Tools Help

Back Search Favorites Media

Address C:\Dream\help12.htm

Go Links

Overview

Getting Started

Working with Map layers
Opening map layers
Viewing labels and legends
Saving map layers
Printing map layers
Closing map layers

Working with Thematic maps
Opening thematic maps
Viewing legends
Saving thematic maps
Printing thematic maps
Closing thematic maps

Working with More Info
Opening More Info
Viewing labels and legends

Overview

SRPIS-Mapper, the mapping module of SRPIS (Sathanur Reservoir Project information System) has been designed specially for GIS requirements. The user can browse through various facets of SRPIS as map layers, thematic features, the attribute data related to these maps, hydraulic designs of Sathanur reservoir, and information on tourist interests in-and-around Sathanur. All the maps have been designed to be self explanatory, with emphasis on clear legends and map scales. However, in SRPIS one can't edit, manipulate, or create new thematic maps. To get around this shortcoming, we have incorporated options for installation of *Map Proviewer 6.5* and downloading of *SRPIS Map Data* from the SRPIS installation CD-ROM disk.

Done My Computer

Fig. 20.18: **The help contents of *SRPIS-Mapper***

SRPIS Map Data

This component of *SRPIS-ACC* can be copied onto the 'C' of the system by following the procedure mentioned in the *Installation Module*.

The *SRPIS Map Data* consists of vector-based digital maps of Sathanur reservoir project. These maps are in the *'.tab'* format which can be read using the *MapInfo ProViewer 6.5* or a professional/run-time version of *MapInfo 6.5* (or later versions). It is also possible to *import* these data formats into other standard GIS software such as *Geomedia* and *Arc Info,* using the *universal translator* option. *Universal translator* enables the user to convert one GIS file format to other GIS file format. Also, we have provided several *workspaces* (a group of organized *'.tab' files*) which can be called in *MapInfo Proviewer/Map Info* (run-time or professional version) for ready reference.

The methodology involved in making the digital maps of SRPIS has been presented as Figure—20.19.

The *SRPIS Map Data* consists of the following maps:

A. India and Tamil Nadu

1. Political map of *India*
2. Political map of Tamil Nadu: (a) districts (b) taluks (c) major cities and (c) capital

B. Ponnaiyar Basin and Catchment

1. Ponnaiyar basin and the catchment of Sathanur reservoir
2. SRP contours: (a) line contours (b) point data (c) grid and (d) digital elevation model

C. SRP Command Area

1. SRP command area
2. SRP streams
3. SRP road-network: with the features of footpath, cart road, pukka road, railways
4. SRP village (polygon)
5. SRP villages (points)

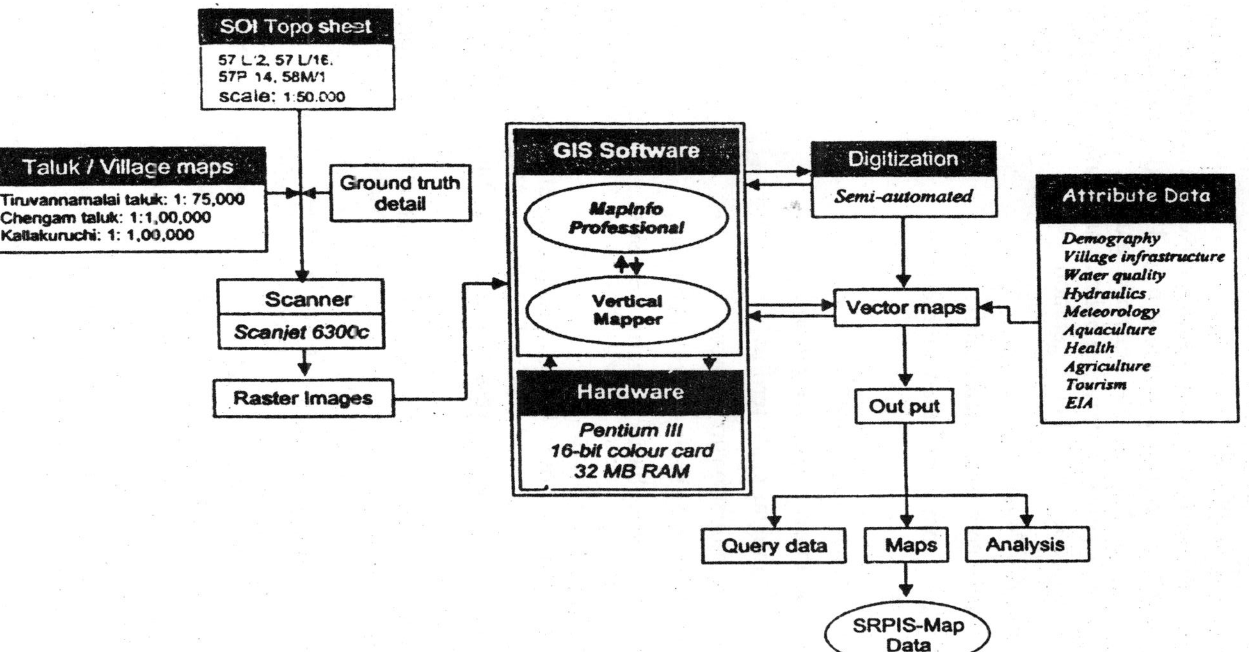

Fig. 20.19: The process and the evolution of *SRPIS Map Data*

6. SRP settlements
7. SRP vegetation: dense scrub, mixed scrub, open mixed scrub, open scrub, dense mixed jungle, open mixed jungle, open jungle, vegetable garden, cashew, vine & trellis, grass, palmyra & palms, casuarinas, eucalyptus, bamboo, and other vegetation.

D. Thematic Maps

1. SRP command area health
2. SRP command area water logged
3. Ponnaiyar basin geology
4. Ponnaiyar basin soils
5. Ponnaiyar basin irrigation
6. Ponnaiyar basin fertility
7. Ponnaiyar basin literacy
8. Ponnaiyar basin sex-ratio general
9. Ponnaiyar basin sex-ratio < 6 years

Map ProVeiwer 6.5

This component of *SRPIS-ACC* can be installed by following the procedure mentioned in the *Installation Module*. For *Map Proviewer* to run properly it is essential that *MrSID* and *PVD65DAO* are installed properly.

Map ProViewer is a simple, inexpensive application that opens, displays, and allows limited manipulation of MapInfo Tables and Workspaces (Figure—20.20). Map ProViewer works with MapInfo Tables and Workspaces created in any version of MapInfo. *Map Proviewer 6.5* is a free-ware and a product of *MapInfo Inc.*

Map ProViewer has been included in the package of *SRPIS* as an alternative to working with paper maps, particularly because:

1. There is a lot of detail and the paper map would have to be very large
2. There are many related maps that would require many pages
3. The user may wish to take professional quality prints of the maps.

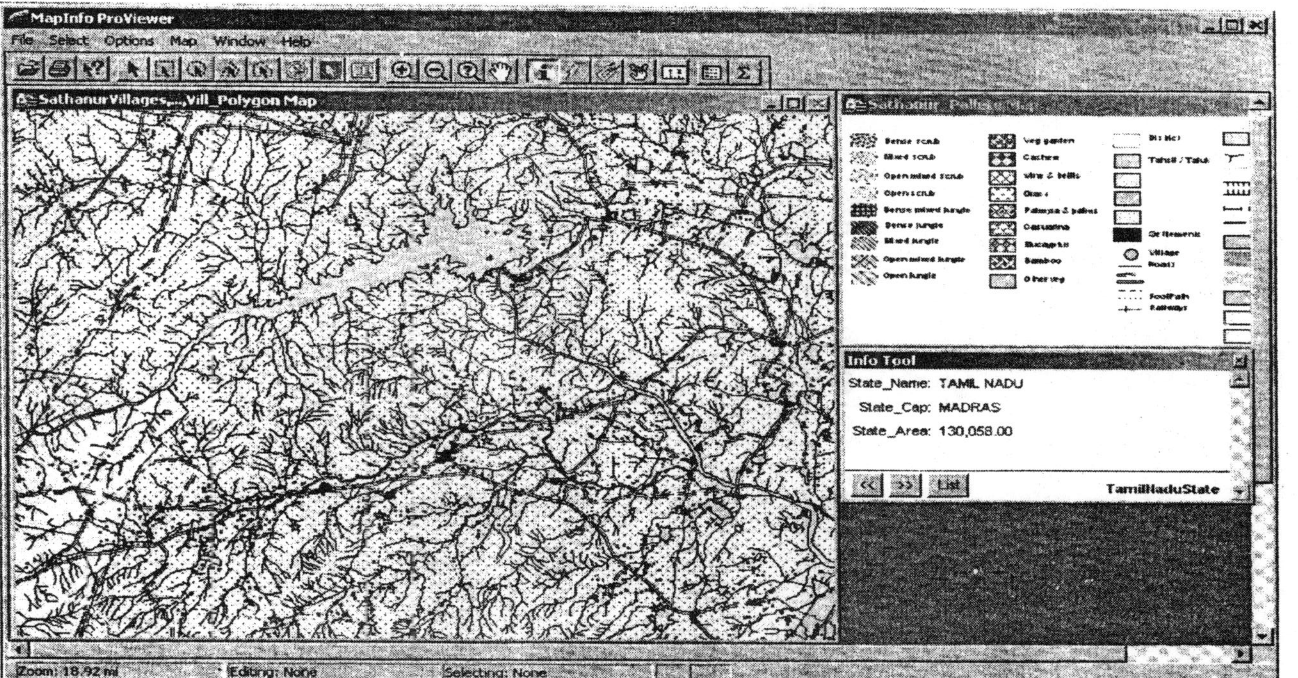

Fig. 20.20: Running the application of *Map ProViewer* using the *SRPIS-DBM*

21

SRPIS Reference Guide

Contents of the SRPIS Database Module *(SRPIS-DBM)*

The *SRPIS-DBM* has the following data:

1. The 1991 census data has been appended with the *SRPIS-DBM* tables under the following groups:
   (a) Demography: (i) *population,* (ii) *social and cultural* (iii) *labour - general* (iv) *labour - details*
   (b) Village infrastructure: (i) *educational institutions* (ii) *medical institutions* (iii) *post and telephone* (iii) *drinking water facilities* (iv) *market* (v) *communication facilities*

2. Water Quality
   (a) *Surface water quality* of Sathanur reservoir and pickup anicut, from 1985 till 2002
   (b) Village-wise *ground water quality* data, from 1972 till 1991

3. Hydraulics
   (a) Daily *inflow-outflow* data of Sathanur reservoir from 1970 till 2001
   (b) Monthly *inflow-outflow* data of Sathanur reservoir from 1977 till 2001

(c) *Max-flood discharge* of Sathanur reservoir, on daily basis, from 1970 till 2001

4. Meteorology

(a) *Temperature*

(i) Daily ambient *temperature* data at Sathanur reservoir from 25.09.1993 till 01.09.1994

(ii) Monthly ambient *temperature* data at Tiruvannamalai, from 2000 till 2001

(b) *Rainfall*

(i) Daily *rainfall* data for right bank canal (RBC) from 01.09.99 till 25.04.01; and for Sathanur reservoir from 01.01.1962 till 31.12.2001

(ii) Monthly *rainfall* data for Sathanur reservoir from 1977 till 2000

5. Aquaculture

(a) Annual *aquaculture* data for Sathanur reservoir from 1978 till 2002

(b) Monthly *aquaculture* data for Sathanur reservoir from 1999 till 2002

(c) *Revenue* generated form aquaculture for Sathanur reservoir from 1977 till 1997

6. Health: village wise health data for Sathanur command area from 1994 till 2000

7. Agriculture

(i) *Irrigation:* 1991 census data

(ii) *Land use:* 1991 census data

(iii) *Cropping pattern*

(a) *Food crops:* village-wise data for command area from 1997 till 1998

(b) *Cash crops:* village-wise data for command area from 1996 till 1999

8. Tourism
   (i) *Monthly:* Presently the monthly data on tourism is not available. This template has been made for use in future.
   (ii) *Annual data* of tourism for Sathanur reservoir from 1991 till 1999.
9. EIA: For the methodology used to generate the database for EIA module, kindly refer *chapter 6* of Volume I.
   (i) *EIA matrix:* The cumulative environmental impact assessment ratings of the observers/evaluators is stored in this table.
   (ii) *EIA personal:* The personal details of the observers/evaluators are stored in this table.

Table design of SRPIS Database Module (SRPIS-DBM)

Table—21.1 Administrative boundaries

Field Name	Field Code	Unit/type	Description
District	District	Text	District name
District code	districtCode	Text	District code
Taluk	Taluk	Text	Taluk name
Taluk code	Talukcode	Text	Taluk code
Village	Village	Text	Village name
Village code	Villagecode	Text	Village code

Table—21.2 Site

Field Name	Field Code	Unit/type	Description
Site	Site_name	Text	Site name
Site code	Site_Code	Text	Site code

Table—21.3 Demography: ***population***

Field Name	Field Code	Unit	Description
Total Pop	T_POPULAT	Raw (no.)	Total population
Density	DENSITY	Inh./km^2	Population density
House Holds	HOUSEHOLDS	Raw (no.)	Number of households
Male pop	MALE_POP	Raw (no.)	Male population
Female Pop	FEMALE_POP	Raw (no.)	Female population
Sex ratio	SR_POPULAT	Female/male	Sex ratio
Total children	T_CHILDREN	Raw(< 6yrs. Nos)	Number of children (below 6 years)
Male children	M_CHILDREN	Raw (no.)	Number of male children (below 6 years)
Female children years)	F_CHILDREN	Raw (no.)	Number of female children (below 6
Sex ratio of children	SR_CHILDREN	Male/Female	Sex ratio (below 6 years)

Table—21.4 Demography: *Social and cultural*

Field Name	Field Code	Unit	Description
Total SC Pop	T_SC	Raw (no.)	Total scheduled caste population (SC)
% SC of Tot Pop	T%_SC	%	% SC population of the total population
Male SC Pop	M_SC	Raw (no.)	Total SC male population
Female SC Pop	F_SC	Raw (no.)	Total SC female population
SC Sex Ratio	SR_SC	Male/Female	Sex ratio of SC
Total ST Pop	T_ST	Raw (no.)	Total scheduled tribe (ST) population
% SC of Tot Pop	T%_ST	Raw (no.)	% ST of the total population
Male ST Pop	M_ST	Raw (no.)	Total male ST population
Female ST Pop	F_ST	Raw (no.)	Total female ST population
ST Sex Ratio	SR_ST	Raw (no.)	Sex ratio of ST
Total LT Pop	T_LITTERAT	Raw (no.)	Total literate population
% LT of Tot Pop	T%_LITTERA	Raw (no.)	% of literate population above 5
Male LT	M_LITTERAT	Raw (no.)	Literate male population
Female LT	F_LITTERAT	Raw (no.)	Literate female population
Sex Ratio LT	SR_LITTERA	Male/Female	Sex ratio of the literate

Table—21.5 Demography: ***labour (general)***

Field Name	Field Code	Unit	Description
Total Wks	T_WORKERS	Raw (no.)	Number of total workers
% Wks in Pop	T%_WORKERS	%	% of workers of the total population
Male Wks	M_WORKERS	Raw (no.)	Number of male workers
Female Wks	F_WORKERS	Raw (no.)	Number of female workers
Sex ratio of Wks	SR_WORKERS	Male/Female	Sex ratio of working population
Total MWks	T_MARGINAL	Raw (no.)	Number of Marginal workers
% MWks in Pop	T%_MARGINA	%	% of marginal workers of the total population
Male MWks	M_MARGINAL	Raw (no.)	Number of male marginal workers
Female MWks	F_MARGINAL	Raw (no.)	Number of female marginal workers
Sex ratio of MWks	SR_MARGINA	Male/Female	Sex ratio of marginal workers
Total NWks	T_NONWORK	Raw (no.)	Total number of non- working population
% NWks in Pop	T%_NONWORK	%	% of non-workers of the total population
Male NWks	M_NONWORK	Raw (no.)	Number of male non- workers
Female NWks	F_NONWORK	Raw (no.)	Number of female non- workers
Sex ratio of NWks	SR_NONWORK	Male/Female	Sex ratio of non-working population
Total Wks_PS	T_PRIMARY	Raw (no.)	Total workers in the primary sector
% Wks_PS in Pop	T%_PRIMARY	%	% of workers in the primary sector of the total population
Male Wks_PS	M_PRIMARY	Raw (no.)	Male workers in the primary sector

(Table Contd...)

1	2		3
Female Wks_PS	F_PRIMARY	Raw (no.)	Female workers in the primary sector
Sex ratio of Wks_PS	SR_PRIMARY	Male/Female	Sex ratio of workers in the primary sector
Total Wks_SS	T_SECUND	Raw (no.)	Total workers in the secondary sector
% Wks_SS in Pop	T%_SECUND	%	% of workers in the secondary sector of the total population
Male Wks_SS	M_SECUND	Raw (no.)	Male workers in the secondary sector
Female Wks_SS	F_SECUND	Raw (no.)	Female workers in the secondary sector
Sex ratio of Wks_SS	SR_SECUND	Male/Female	Sex ratio of workers in the secondary sector
Total Wks_SS	T_TERTIARY	Raw (no.)	Workers in the tertiary sector
% Wks_SS in Pop	T%_TERTIAR	%	% of workers in the tertiary sector of the total population
Male Wks_SS	M_TERTIARY	Raw (no.)	Male workers of the tertiary sector
Female Wks_SS	F_TERTIARY	Raw (no.)	Female workers of the tertiary sector
Sex ratio of Wks_SS	SR_TERTIAR	Male/Female	Sex ratio of the workers in the tertiary sector

Table—21.6 Demography: *labour (details)*

Field Name	Field Code	Unit	Description
Total CL	T_CULTIVAT	Raw (no.)	Cultivators
% of CL	T%_CULTIVA	% (no.)	% of cultivators
Male CL	M_CULTIVAT	Raw (no.)	Male cultivators
Female CL	F_CULTIVAT	Raw (no.)	Female cultivators
Sex ratio of CL	SR_CULTIVA	Male/female (no.)	Sex ratio of the cultivators
Total AL	T_AGRI_LAB	Raw (no.)	Total agricultural labourers
% of AL	T%_AGRI_LA	% (no.)	% of agricultural labourers of the total population
Male AL	M_AGRI_LAB	Raw (no.)	Male agricultural labourers
Female AL	F_AGRI_LAB	Raw (no.)	Female agricultural labourers
Sex ratio of AL	SR_AGRI_LAB	Male/female (no.)	Sex ratio of the agricultural labourers
Total FF	T_FISH_FOR	Raw (no.)	Total population of the fishing community
% of FF	T%_FISH_FO	% (no.)	% of fishing population of the total population
Male FF	M_FISH_FOR	Raw (no.)	Male fishing population
Female FF	F_FISH_FOR	Raw (no.)	Female fishing population
Sex ratio of FF	SR_FISH_FO	Male/female (no.)	Sex ratio of the fishing population
Total Mi	T_MINING	Raw (no.)	Total population depending on mining and quarrying
% of Mi	T%_MINING	% (no.)	% of population depending on mining and quarrying of the total population
Male Mi	M_MINING	Raw (no.)	Male mining and quarrying workers

(Table Contd...)

1	2	3	4
Female Mi	F_MINING	Raw (no.)	Female mining and quarrying workers
Sex ratio of Mi	SR_MINING	Male/female (no.)	Sex ratio of mining and quarrying population
Total HHI	T_HH_INDUS	Raw (no.)	Total manufacturing /processing industry workers in household
% of HHI	T%_HH_INDU	% (no.)	% of manufacturing /processing in household industry workers of the total population
Male HHI	M_HH_INDUS	Raw (no.)	Male manufacturing / processing in household industry workers
Female HHI	F_HH_INDUS	Raw (no.)	Female manufacturing /processing in household industry workers
Sex ratio of HHI	SR_HH_INDU	Male/female (no.)	Sex ratio of household industry workers
Total OI	T_OTH_INDU	Raw (no.)	Total manufacturing /processing (non-household industry) workers
% of OI	T%_OTH_IND	% (no.)	% of manufacturing /processing (non-household industry) workers of the total population
Male OI	M_OTH_INDU	Raw (no.)	Male Manufacturing /processing (non-household industry) male workers
Female OI	F_OTH_INDU	Raw (no.)	Female manufacturing /processing (non-household industry) female workers
Sex ratio of OI	SR_OTH_IND	Male/female (no.)	Sex ratio of manufacturing /processing (non-household industry) workers
Total Con	T_CONSTRUC	Raw (no.)	Total population of the construction workers
% of Con	T%_CONSTRU	% (no.)	% of construction workers of the total population
Male Con	M_CONSTRUC	Raw (no.)	Male construction workers
Female Con	F_CONSTRUC	Raw (no.)	Female construction workers

(Table Contd...)

1	2	3	4
Sex ratio of Con	SR_CONSTRU	Male/female (no.)	Sex ratio of construction workers
Total Trd	T_TRADE	Raw (no.)	Trade and commerce workers
% of Trd	T%_TRADE	% (no.)	% of Trade and commerce workers of the total population
Male Trd	M_TRADE	Raw (no.)	Male trade and commerce workers
Female Trd	F_TRADE	Raw (no.)	Female trade and commerce workers
Sex ratio of Trd	SR_TRADE	Male/female (no.)	Sex ratio of trade and commerce workers
Total Tra	T_TRANSPOR	Raw (no.)	Total transportation, storage and communication workers
% of Tra	T%_TRANSPO	% (no.)	% of Transportation, storage and communication workers of the total population
Male Tra	M_TRANSPOR	Raw (no.)	Male transportation, storage and communication workers
Female Tra	F_TRANSPOR	Raw (no.)	Female Transport, storage and communication workers
Sex ratio of Tra	SR_TRANSPO	Male/female (no.)	Sex ratio of transport, storage and communication workers
Total Ser	T_SERVICES	Raw (no.)	Total other services workers of the tertiary sector
% of Ser	T%_SERVICE	% (no.)	% of other services workers of the tertiary sector of the total population
Male Ser	M_SERVICES	Raw (no.)	Other services male workers of the tertiary sector
Female Ser	F_SERVICES	Raw (no.)	Other services female workers of the tertiary sector
Sex ratio of Ser	SR_SERVICE	Male/female (no.)	Sex ratio of other services workers of the tertiary sector

Table—21.7 Village infrastructure: *educational institutions*

Field Name	Field Code	Unit	Description
Availability	EI_AVALAIB	Yes/no	Whether educational institutions are available
Nearest EI	EI_DISTANC	Distance class	Distance to the nearest educational institution (as distance class: <5Km, 5-10 Km, >10 Km))
Primary schools	P_SCHOOL	Number (no.)	Number of Primary schools
Middle schools	M_SCHOOL	Number (no.)	Number of middle schools
High Schools_School	H_SCHOOL	Number (no.)	Number of high schools
PU Schools	PU_COLLEGE	Number (no.)	Number of PU schools
Graduate Colleges	GR_COLLEGE	Number (no.)	Number of graduate colleges
Adult Literacy Centres	ADULT_LIT_	Number (no.)	Number of adult literacy centres
Industrial schools	INDUS_SCH	Number (no.)	Number of industrial schools
Training Schools	TRAIN_SCH	Number (no.)	Number of training schools
Other EI	OTHER_EI	Number (no.)	Number of other educational institutions

Table—21.8. Village infrastructure: *medical institutions*

Field Name	Field Code	Unit	Description
Availability	MI_AVALAIB	Yes/no	Whether medical institutions are available
Nearest MI	MI_DISTANC	Distance class	Nearest medical institution (as distance class: <5Km, 5-10Km, >10 Km)
Hospital	HOSPITAL	Number (no.)	Number of hospitals
Maternity Welfare Centre	MAT&CHI_WC	Number (no.)	Number of maternal and child welfare centres
Maternity Homes	MATERNITY	Number (no.)	Number of maternity homes
Child Welfare Centre	CHILD_W_C	Number (no.)	Number of child welfare centres
Primary Health Centre	PRIM_H_C	Number (no.)	Number of primary health centres
Dispersary	DISPENSARY	Number (no.)	Number of dispensaries
Family Planning Centre	FP_C	Number (no.)	Number of family planning centres
Tuberculosis Centre	TB_C	Number (no.)	Number of tuberculosis clinics
Nursing Home	NURSING_H	Number (no.)	Number of nursing homes
Community health workers Centre	CHWC	Number (no.)	Number of community health worker centres
Registered private practitioner	PRIV_PRACT	Number (no.)	Number of registered private practitioners
Subsidiary medical practitioner	SUBS_PRACT	Number (no.)	Number of subsidiary medical practitioners
Other Medical centres	OTHER_MI	Number (no.)	Number of other medical centres

Table—21.9 Village infrastructure: post and telephone

Field Name	Field Code	Unit	Description
Availability	PT_AVALAIB	Yes/no	Whether post and telephone facilities available
Nearest Facility	PR_DISTANC	Distance class	Nearest distance to the post and telephone facility (as distance class: <5 Km, 5-10Km, >10Km)
Post Office	POST_OFFIC	Number (no.)	Number of post offices
Telegraph Office	TELEG_OFFI	Number (no.)	Number of telegraph offices
Post and Telegraph Office	PT_OFFICE	Number (no.)	Number of post and telegraph offices
Telephone Connection	PHONE_CONN	Number (no.)	Number of telephone connections

Table—21.10 Village infrastructure: *drinking water facility*

Field Name	Field Code	Type	Description
DWF Availability	DW_AVALAIB	Yes/no	Whether drinking water facility is available
Nearest DWF	DW_DISTANC	distance class	Nearest distance to the drinking water facility (as distance class: <5 Km, 5-10 Km, >10 Km)
Tap Water	DW_TAP	Number (no.)	Number of water taps
Well Water	DW_WELL	Number (no.)	Number of wells
Tube Well Water	DW_TUBEWEL	Number (no.)	Number of tube wells
Hand Pump Water	DW_HANDPUM	Number (no.)	Number of hand pumps
River Water	DW_RIVER	Number (no.)	Number of river water facilities
Fountain Water	DW_FOUNTAI	Number (no.)	Number of fountain water facilities
Spring Water	DW_SPRING	Number (no.)	Number of spring water facilities
Canal Water	DW_CANAL	Number (no.)	Number of canal water facilities
Tank Water	DW_TANK	Number (no.)	Number of tank and lake water facilities
Others	DW_OTHER	Number (no.)	Number of other drinking water sources

Table—21.11 Village infrastructure: *market*

Field Name	Field Code	Type	Description
Availability	MK_AVALAIB	Yes/no	Whether market facility is available
Nearest MK	MK_DISTANC	Number (no.)	Nearest distance to the market (as distance class: <5 Km, 5-10 Km, >10 Km)
Daily MK	DAILY_MK	Number (no.)	Number of daily markets

Table—21.12 Village infrastructure: ***communication facilities***

Field Name	Field Code	Unit	Description
Availability	CF_AVALAIB	Yes/no	Whether communication facilities are available
Nearest CF	CF_DISTANC	Distance class	Nearest communication facility (as distance class: <5 Km, 5-10 Km, >10 Km)
Bus Stop	BUS_STOP	Number (no.)	Number of bus stops
Railway	RAILWAY	Number (no.)	Number of railway stations
Pucca road	PUCCA_ROAD	Number (no.)	Number of pucca roads
Kachha road	KACHCHA_RO	Number (no.)	Number of kachcha roads
Footpath	FOOTPATH	Number (no.)	Number of footpaths
Nearest town	NEAR_TOWN	name	Name of the nearest town
Distance to town	DISST_TOWN	Distance class (Km)	Distance to the nearest town (as distance class (<5Km, 5-10 Km, >10 Km)

Table—21.13 Village infrastructure: ***power supply***

Field Name	Field Code	Unit	Description
Availability	PW_AVALAIB	Yes/no	Whether power supply is available
Domestic PW	PW_DOMESTI	binary	Power for domestic sector
Agriculture PW	PW_AGRICUL	binary	Power for agriculture sector
PW for Industrial/ Commercial	PW_IND_COM	binary	Power for industrial and commercial sector
PW for all purposes	PW_FOR_ALL	binary	Power for all purposes

Table—21.14 Water quality: *surface water*

Field Name	Field Code	Unit	Description
Colour	color	text	Colour of the water
Smell	smell	text	Smell of the water
Turbidity	turb	Number (NTU)	Turbidity of the water
pH	ph	Number (pH units)	Ph value of the water
EC	ec	Number(μ mhos)	Electrical conductivity of the water
Alkalinity	alk	Number (mg/l)	Alkalinity of the water
Total Hardness	th	Number (mg/l)	Total Hardness of the water
Ca Hardness	cah	Number (mg/l)	Calcium hardness of the water
Mg	mg	Number (mg/l)	Magnesium content of the water
Cl	cl	Number (mg/l)	Chloride content of the water
Total Solids	ts	Number (mg/l)	Total solids of the water
TDS	tds	Number (mg/l)	Total dissolved solids of the water
SS	ss	Number (mg/l)	Suspended solids of the water
NO_2	no_2	Number (mg/l)	Nitrite content of the water
NH_3	nh_3	Number (mg/l)	Ammonia content of the water
NO_3	no_3	Number (mg/l)	Nitrate content of the water
PO_4	po_4	Number (mg/l)	Phosphate content of the water
SO_4	so_4	Number (mg/l)	Sulphate content of the water
DO	do	Number (mg/l)	Dissolved oxygen of the water
Fe	fe	Number (mg/l)	Iron content of the water
Fl	fl	Number (mg/l)	Fluoride content of the water
Mn	mn	Number (mg/l)	Manganese content of the water
Water Level	level	Number (m)	Water level of the reservoir

Table—21.15 Water quality: ***ground water***

Field Name	Field Code	Unit	Description
pH	ph	Number (ph units)	pH of the water
EC	ec	Number (μ mhos)	Electrical conductivity of the water
Ca	cah	Number (mg/l)	Calcium hardness of the water
Mg	mg	Number (mg/l)	Magnesium content of the water
K	K	Number (mg/l)	Potassium content of the water
HCO_3	hco_3	Number (mg/l)	Bicarbonate content of the water
CO_3	co_3	Number (mg/l)	Carbonate content of the water
SO_4	so_4	Number (mg/l)	Sulphate content of the water
NO_3	no_3	Number (mg/l)	Nitrate content of the water
TDS	tds	Number (mg/l)	Total dissolved solids of the water
TH	th	Number (mg/l)	Total hardness of the water
RSC	rsc	Number (mg/l)	Resonance content of the water
SAR	sar	Number (mg/l)	SAR value of the water
GCT	gct	Number (mg/l)	GCT value of the water
Na	na	Number (mg/l)	Sodium content of the water
Cl	cl	Number (mg/l)	Chloride content of the water

Table—21.16 Hydraulics: *inflow/outflow (daily)*

Field Name	Field Code	Unit	Description
Inflow	inflow	Number (CS)	Inflow of water into the Reservoir
Outflow	outflow	Number (CS)	Outflow of water from the Reservoir

Table—21.17 Hydraulics: *inflow/outflow (monthly)*

Field Name	Field Code	Type	Description
Inflow	inflow	Number (CS)	Inflow of water into the Reservoir
Outflow	outflow	Number (CS)	Outflow of water from the Reservoir

Table—21.18. Hydraulics: *maximum flood discharge*

Field Name	Field Code	Type	Description
Reservoir Level	level	Number (Ft)	Water level of the reservoir
Reservoir Capacity	capacity	Number (McFt)	Capacity of the reservoir
Surplus Discharge	surplus	Number (cusecs)	Surplus discharge of the reservoir

Table—21.19 Meteorology: *temperature-daily*

Field Name	Field Code	Unit	Description
Mean Min	MeanMin	Number (°C)	Mean minimum temperature
Mean Max	MeanMax	Number (°C)	Mean maximum temperature
Actual Humidity	Humidity Actual	Number (%)	Actual humidity

Table—21.20 Meteorology: *temperature-monthly*

Field Name	Field Code	Unit	Description
Mean Min	MeanMin	Number (°C)	Mean minimum temperature
Mean Max	MeanMax	Number (°C)	Mean maximum temperature
Actual Humidity	Humidity Actual	Number (%)	Actual humidity

Table—21.21 Meteorology: *rainfall-daily*

Field Name	Field Code	Unit	Description
Avg Rainfall	AvgRainfall	Number (mm)	Average rainfall

Table—21.22 Meteorology: *rainfall-monthly*

Field Name	Field Code	Unit	Description
Rainfall-Avg	AvgRainfall	Number (mm)	Average Rainfall

Table—21.23 Aquaculture *(annual)*

Field Name	Field Code	Unit	Description
Species	species	Number	Name of the fish (species)
No's	no	Number	Number of fish (each specie)
Biomass	biomass	Number (avg, Kg)	Average biomass of the fish (each specie)

Table—21.24 Aquaculture *(monthly)*

Field Name	Field Code	Unit	Description
Species	species	Number	Name of the fish (species)
No's	no	Number	Number of fish (each specie)
Biomass	biomass	Number (avg,Kg)	Average biomass of the fish (each specie)

Table—21.25 Aquaculture *(revenue)*

Field Name	Field Code	Unit	Description
Total Fish Yield	totalfishyield	Number (tonnes)	Total fish yield
Fish Catch Revenue	fishrevenue	Number (Rs.)	Fish catch revenue
Fish Seed Production	fishseed _tons	Number (tonnes)	Fish seed production (yield)
Fish Seed Production	fishseed _rs	Number (Rs.)	Fish seed production (revenue)

Table—21.26 Health

Field Name	Field Code	Unit	Description
Indicative Population	indicative population	Number	Indicative population
Malarial Incidents	malarialincidents	Number	Whether native imported/neither
Filirial Incidents	filarialincidents	Number	Whether native imported/neither

Table—21.27 Agriculture: *irrigation*

Field Name	Field Code	Unit	Description
Tank Irrigation	Tank	Hectares (ha)	Hectares irrigated by tanks
Well Irrigation	Well	Hectares (ha)	Hectares irrigated by wells
Canal Irrigation	Canal	Hectares (ha)	Hectares irrigated by canals
Others	Others	Hectares (ha)	Hectares irrigated by other sources
Total Irrigated Area	Total	Hectares (ha)	Hectares of total irrigated area

Table—21.28 Agriculture: *land use*

Field Name	Field Code	Unit	Description
Forest	LU_FOREST	Hectares (ha)	Land use under forest
% Forest of total area	LU%_FOREST	% (ha)	Land use under forest as % of total land area
Total irrigated area	LU_IRRIGAT	Hectares (ha)	Total irrigated area
% Irrigated area	LU%_IRRIGA	%	Irrigated area as % of total cultivated area
Unirrigated area	LU_UNIRRIG	Hectares (ha)	Un-irrigated area
% of Unirrigated area	LU%_UNIRRI	%	Un-irrigated area as % of total cultivated area
Culturable waste	LU_CULT_WA	Hectares (ha)	Culturable waste
% of Culturable waste	LU%_CULT_W	%	Culturable waste as % of total area
NA for cultivation	LU_NON_CUL	Hectares (ha)	Land not available for cultivation
% NA for cultivation	LU%_NON_C	% (ha)	Land not available for cultivation as % of total area

Table—21.29 Agriculture: *cropping pattern (food crops)*

Field Name	Field Code	Unit	Description
Sornavari	sornavari	Hectares (ha)	Land cultivated for *sornavari* rice production
Samba	samba	Hectares (ha)	Land cultivated for *samba* rice production
Navarai	navarai	Hectares (ha)	Land cultivated for *narvarai* rice production
Others	rice_others	Hectares (ha)	Land cultivated for other varieties of rice production
Cumbu	cumbu	Hectares (ha)	Land cultivated for *cumbu* production
Corn	corn	Hectares (ha)	Land cultivated for *corn* production
Ragi	ragi	Hectares (ha)	Land cultivated for *ragi* production
Others	cereals_others	Hectares (ha)	Land cultivated for other food crops
Red Gram	redgram	Hectares (ha)	Land cultivated for *red gram* production
Black Gram	blackgram	Hectares (ha)	Land cultivated for *black gram* production
Horse Gram	horsegram	Hectares (ha)	Land cultivated for *horse gram* production
Green Gram	greengram	Hectares (ha)	Land cultivated for *green gram* production
Others	pulses_others	Hectares (ha)	Land cultivated for other pulses production

Table—21.30 Agriculture: *cropping pattern (cash crops)*

Field Name	Field Code	Unit	Description
Turmeric	turmeric	Hectares (ha)	Land cultivated for *turmeric* production
Mustard	mus	Hectares (ha)	Land cultivated for *mustard* production
Chilly	chilly	Hectares (ha)	Land cultivated for *chilly* production
Others	spices_others	Hectares (ha)	Land cultivated for other cash crops production
Brinjal	brinjal	Hectares (ha)	Land cultivated for *brinjal* production
Potato	potato	Hectares (ha)	Land cultivated for *potato* production
Onion	onion	Hectares (ha)	Land cultivated for *onion* production
Mango	mango	Hectares (ha)	Land cultivated for *mango* production
Tomato	tomato	Hectares (ha)	Land cultivated for *tomato* production
Bhendi	bhendi	Hectares (ha)	Land cultivated for *bhendi* production
Pumpkin	pumpkin	Hectares (ha)	Land cultivated for *pumpkin* production
Others	fruits_others	Hectares (ha)	Land cultivated for other fruits production
Groundnut	ground nut	Hectares (ha)	Land cultivated for *ground nut* production
Coconut	coconut	Hectares (ha)	Land cultivated for *coconut* production
Gingelly	cingelly	Hectares (ha)	Land cultivated for *gingelly* production
Castor	castor	Hectares (ha)	Land cultivated for *castor* production
Others	oil_others	Hectares (ha)	Land cultivated for other oil crops
Sugarcane	sugarcane	Hectares (ha)	Land cultivated for *sugarcane* production
Cotton	cotton	Hectares (ha)	Land cultivated for *cotton* production
Plantain	plantain	Hectares (ha)	Land cultivated for *plantain* production
Palm	palm	Hectares (ha)	Land cultivated for *palm* production
Others	crops_others	Hectares (ha)	Land cultivated for other crops production

Table—21.31 Tourism: *annual*

Field Name	Field Code	Unit	Description
Tourist Inflow	touristinflow	Number (no.)	Number of tourists visited
Revenue	revenue	Number (Rs.)	Revenue generated

Table—21.32 Tourism: *monthly*

Field Name	Field Code	Unit	Description
Tourist Inflow	touristinflow	Number (no.)	Number of tourists visited
Revenue	revenue	Number (Rs.)	Revenue generated

Table—21.33 EIA: *personal details*

Field Name	Field Code	Unit/ type	Description
Village name	Villagename	Text	Name of the village for which the evaluation is being made
Village code	Villagecode	Number	Code of the village
Location	Location	Text	Whether the village is in SLBC or SRBC
Category	Category	Text	Whether the observer is an *end user, expert*, or *other*
Person	Person	Text	Name of the observer
Age	Age	Number	Age of the observer
Occupation	Occupation	Text	Occupation of the observer
Education	Education	Text	Education of the observer
Income	Income	Number	Income of the observer
Holding	Holding	Number	The area of land holding, if the evaluator is an *enduser*.

Table—21.34 EIA: *matrix*

Field Name	Field Code	Unit/type	Description
Village name	Villagename	Text	Name of the village for which the evaluation is being made
Village code	Villagecode	Number	Code of the village
Location	Location	Text	Whether the village is in SLBC or SRBC
Category	Category	Text	Whether the evaluator is *end user*, *expert*, or *other*
Area brought under cultivation	Ditto as *field name*	Number	Ditto as *field name*
Productivity per hectare	Ditto as *field name*	Number	Ditto as *field name*
Usage of Chemical fertilizers	Ditto as *field name*	Number	Ditto as *field name*
Usage of bio fertilizers	Ditto as *field name*	Number	Ditto as *field name*
Problem of weeds in the fields	Ditto as *field name*	Number	Ditto as *field name*
Pests menace in the crops	Ditto as *field name*	Number	Ditto as *field name*
Usage of weedicides and pesticides	Ditto as *field name*	Number	Ditto as *field name*
Scientific approach towards agriculture	Ditto as *field name*	Number	Ditto as *field name*
Agriculture socities and Bank	Ditto as *field name*	Number	Ditto as *field name*
Source of irrigation	Ditto as *field name*	Number	Ditto as field name
Mode of irrigation	Ditto as *field name*	Number	Ditto as *field name*

(Table Contd...)

1	2	3	4
Water logging in the fields	Ditto as *field name*	Number	Ditto as *field name*
Salinity and alkanity in the field	Ditto as *field name*	Number	Ditto as *field name*
Water logging in the fields near to canal	Ditto as *field name*	Number	Ditto as *field name*
Canal irrigation / tank irrigation	Ditto as *field name*	Number	Ditto as *field name*
Modern implements of agriculture and irrigation	Ditto as *field name*	Number	Ditto as *field name*
Drainage in the fields	Ditto as *field name*	Number	Ditto as *field name*
Lining of the canal	Ditto as *field name*	Number	Ditto as *field name*
Vegetation in the vicinity of canal	Ditto as *field name*	Number	Ditto as *field name*
Vegetation in the vicinity of tanks	Ditto as *field name*	Number	Ditto as *field name*
Siltation in the canal	Ditto as *field name*	Number	Ditto as *field name*
Siltation in the tank	Ditto as *field name*	Number	Ditto as *field name*
Seepage from canal	Ditto as *field name*	Number	Ditto as *field name*
Optimal flow of water from reservoir through canals	Ditto as *field name*	Number	Ditto as *field name*
Judious distribution of water to the fields	Ditto as *field name*	Number	Ditto as field name
Breaching and enroachment in the canal and tanks	Ditto as *field name*	Number	Ditto as *field name*
Income	Ditto as *field name*	Number	Income of the observer
Primary heath centres	Ditto as *field name*	Number	Ditto as *field name*
Water borne disease	Ditto as *field name*	Number	Ditto as *field name*
Schools	Ditto as *field name*	Number	Ditto as *field name*

(Table Contd...)

1	2	3	4
Forest	Ditto as *field name*	Number	Ditto as *field name*
Roads	Ditto as *field name*	Number	Ditto as *field name*
Drinking water	Ditto as *field name*	Number	Ditto as *field name*
Power supply	Ditto as *field name*	Number	Ditto as *field name*
Employment and migration	Ditto as *field name*	Number	Ditto as *field name*

22

SRPIS in Future

"*Change is the only thing which is constant*"- this phrase, which applies to all aspects of life, does so even more poignantly to computer software and hardware. However 'perfect looking' a software may be, it soon gives way to something better.

We envisage that the evolution of SRPIS would occur along the lines described below:

1. As of now it is not possible to view different maps all at once in the restricted space of the map browser designed for SRPIS. This facility can be added in future by enabling the software to open different maps in different browser windows. This would be achievable after some minor modifications in the *standard windows* features are affected.

2. Several combinations of map layers would be desirable, and often essential, for a very large GIS project. Presently, in SRPIS, so also in most of the web-based GIS, the maps are loaded as 'images' in the form of GIF or JPEGS. As the number of layers increase, the possible combinations of the layers also increase, which means the software needs to handle more images. This increase in the images burdens the server and substantially increases the time for loading and displaying of

information. Incidentally this is one of the major drawbacks of all web-based GIS packages thus far.

The authors are currently striving to overcome this general shortcoming of web-based GIS tools by designing a new system of handling map objects in such environments.

PART—C

Conclusions and Recommendations

23

Conclusions and Recommendations

Impacts, Upstream

Considering the geomorphology and the land use/land cover pattern, catchment management deserves utmost priority. This would curtail a most of the negative impacts presently occurring due to the of Sathanur Reservoir Project.

Thus far, some silt retention dams have been constructed in the catchment. However, construction of contour bunds at regular intervals and at appropriate places, based on the terrain slope and soil quality would yield better results. Check dams need be constructed at places where ravines and gullies occur. While the contour bunds if constructed scientifically, would conserve the top soil which is much needed for the growth of ground cover, the contour bunds would also help in recharging the aquifers and arresting soil erosion, which in turn would reduce sedimentation of the reservoir.

Because of the rapid soil erosion in this region, it is necessary to survey the slope contours for the SRP catchment and command areas with immediate effect.

Mass awareness among the people is needed for conserving the existing forestry, and to take up afforestation programmes on a large scale and at regular intervals.

We urge that the forest officials, with the help of agriculture department, take up a study to identify the plants that can be grown in the catchment and command areas of SRP, based on the soil quality and the prevailing environmental conditions.

IMPACTS ON THE RESERVOIR AND CATCHMENT

Inundation, Seismicity and Hydroelectricity

Sathanur reservoir is constructed in a forest with almost no human habitation. Hence the human displacement due to inundation was practically nil. Also, the forest land in-and-around the reservoir doesn't support any unique or rare plants which have been categorized as endangered. In an attempt to compensate for the forest was cleared for the construction of the reservoir, trees have been planted elsewhere.

The past experience indicates that several reservoirs across the world have induced earthquakes. Thus, siesmicity is assessed routinely for most of the reservoir projects and this has become an integral part of the standard environmental impact assessment (EIA) and environmental management planning.

In the case of SRP, neither prior to the construction of the dam nor after the construction of the dam, seismic studies have been carried out. It is therefore recommended that steps may be taken for monitoring of the seismicity of this region on a regular basis. The information thus generated would be useful not only in assessing the environmental impacts of Sathanur Reservoir project but also in its conservation and management.

The cause of siltation in Sathanur Reservoir has been attributed to land use and land cover pattern in the SRP catchment area. Poor soil holding capacity, sparse ground cover and harsh weather combine to worsen the problem.

Some of the ecorestoration measures that were undertaken to abate soil erosion are sediment traps or check dams, and afforestation in the catchment area of SRP. As mentioned earlier, we feel, construction of contour bunds in the catchment of SRP based on the terrain contours can work wonders by allowing seepage of surface water. It would help in the growth of ground cover as well as recharging the aquifers. The growth of ground cover would arrest soil erosion which, in turn, would reduce sedimentation in the Sathanur reservoir.

Hydro electricity is one of the clean and environment friendly energy sources when compared with the use of fossil fuels. The hydro-electricity production unit of SRP caters to a mere 0.2 per cent of the total production of electricity and 0.8 per cent of the hydro-electricity production in the state. However, the profit incurred by the SRP due to the hydro-electricity production is quite good.

From the existing information, it is not clear whether the present production capacity of the SRP hydro-electricity unit is the maximum and whether there is still any scope of enhancing the production capacity. Perhaps, a study might be conducted to look into the feasibility of enhancing the production of hydroelectricity, so that the benefits from the SRP can be maximized.

Recreation

Sathanur reservoir is a popular tourist destination for the people living in the nearby towns of Thiruvannamalai, Thirupatthur, Cuddalore, Villupuram, Pondicherry, and Chennai.

Tourism in Sathanur reservoir has earned far more revenue than the expenditure incurred in developing and maintaining it.

The main attractions of Sathanur as a tourist spot include: (i) scenic beauty (ii) idyllic surroundings which create an amblence of tranquility and peace (iii) a crocodile farm (iv) a swimming pool (v) a mini zoo and (vi) an aquarium of ornamental fishes.

As of now, the tourist inflow hasn't generated discernable adverse impacts. This is principally due to the care and attention bestowed by the reservoir authorities on the collection and treatment of wastes generated by the tourists.

Even though the Sathanur Reservoir Project (SRP) is a net revenue earner vis-a-vis tourism, there appears to be much greater potential of developing it as an eco tourism spot than realized so far. Based on our studies, and feedback from the authorities, the following guidelines emerge:

(a) A balance should be struck between attracting tourists and maintaining the ecological well-being of SRP. In view of the depth of the reservoir and presence of crocodiles, manual boating may be avoided but motor-boats taking tourists around the reservoir and to an island within the reservoir can be contemplated.

(b) On-land facilities for sport and pastime in the backdrop of the reservoir can be substantially increased so that a visitor can spend a few days at the site.

(c) 'Tree houses' and 'meditation chambers' may be created especially for the foreigners. These have good potential for earning revenue without unduly stressing the environment.

(d) Simultaneously, facilities for transportation, boarding and lodging may be created partly through governmental resources and partly through private entrepreneurship.

In summary, if developed in an ecologically viable manner, SRP shall become not only a much bigger tourist attraction but also a source of employment and commerce for the people of the region.

Health

Inspite of the best efforts of the officers, associated with the command area of water resources projects, to prevent water-logging, the latter does occur often. The main reservoir, the canals, and the water-logged land become breeding niches for pests leading to outbreaks of malaria, filaria, and schistosomiasis. Water-logging can also lead to anoxic conditions at the water-soil interface and cause luxury release of metals such as aluminum, manganese and molybdenum which are harmful to plants and animals.

The swampy conditions prevailing in heavily water-logged areas can be particularly dangerous as such areas are also rich in animal excreta with the support of which mosquitoes and other disease vectors can multiply with alarming fluency.

So far no case of filaria has been reported at SRP. There were several incidences of malaria during the years 1994-95, when the construction work on the hydro-electricity unit had began. But there had been no increase in the area covered with water due to the construction activity and investigations led to the finding that the malaria was not due to SRP but was rather 'imported'—it was brought from outside to the project site by workers infected with the disease. Indeed the annual parasite index at SRP has been steadily declining since 1944, barring the exception of 1994-95.

As of now the infrastructural development induced by the SRP, such as better roads and transportation facilities, have lead to an increase in the inflow of hired labourers in this region. This has introduced a potential risk of diseases being brought to the SRP from outside.

The following mitigation measures are being taken by the officials to control the spread of malaria:

- Periodic blood smear tests are performed on the work force employed at the Sathanur dam site.
- The labourers from the nearby regions aspiring to be hired are screened for any unusual symptoms before they are accepted.
- Anti-malarial drugs (mostly chloroquines) are distributed free of cost to the residents at the dam site.
- DDT, BHC and other chemicals are sprayed regularly indoors and outdoors at the dam area to check the menace of mosquitoes.
- The mosquito larvae in the ponds and drains are killed using various larvicide formulations.
- Burrow pits, soakage pits and other semi-permanent water pools in the forest, hills and in the vicinity of the reservoir are periodically monitored and closed.

The villages downstream SRBC/SRBC have been classified into three categories based on the annual parasite index (API) values. The intensity of monitoring and the mitigation measures to be taken are decided on the basis of the API range under which the given village falls.

Aquaculture

The annual catch of the ichthyofauna of Sathanur reservoir has displayed a wide range of fluctuation over a span of twenty years (1978-98). Except for *Labeo rohita, Labeo calabasu* and some cat-fishes, the yields of all other fishes have declined over the past twenty years. Yields of *Cyprinus carpio* and *Cyprinus cirrhosa* have been reduced to zero.

We found no significant correlation between the reservoir hydrology—annual precipitation, inflow, outflow and water level—and the annual catch of the ichthyofauna. At some points of time these do seem to influence the annual catch but there is no uniform pattern.

The zoomass per fish of *Catla catla* reduced from 9.7 kg in 1979-80 to 2.9 kg in 1998, while that of *Labeo rohita* plunged to 0.3 kg in 1998 from 1.7 kg in 1978. The zoomass of *Cyprinus carpio* reduced to 1.5 - 1.6 (1997-98) from 4.9 kg in 1979-80. The zoomass of *Cyprinus cirrhosa* came down to 0.9 kg in 1998 from 2 kg in 1986. *Cirrhinus mrigal, Labeo fimbriatus* and catfishes maintained almost steady zoomass during the study period.

The Sathanur reservoir waters are currently dominated by *Catla catla, Labeo rohita, Cirrhinus mrigal* and *Labeo calabasu*, in that order, in terms of total catch. The catch of other species-*Labeo fimbriatus, Cyprinus cirrhosa, Wallago attu*, cat-fishes and others have declined considerably, while that of *Cyprinus cirrhosa* being reduced to zero. The transplanted species of fishes have, thus, established themselves at the expense of the indigenous inhabitants.

The total catch of the reservoir has come down considerably from 237 tons in 1981 to 151 tons in 1998. Water quality of the reservoir, in terms of high pH, alkalinity, hardness, turbidity and total solids concentration, may have had a profound impact on the fisheries. Low concentration of dissolved oxygen in the reservoir may also have a strong bearing on the fish catch.

The brighter side of the Sathanur aquaculture is that annual revenue and profits have steadily increased. This is because the market value of the catch has gone up even as the quantities of the catch have declined. It also indicates that fisheries are another aspect of Sathanur. reservoir, which has much greater potential than harnessed thus far.

Certain key water quality parameters, such as pH, DO, and TDS, which are known to provide clue to the health of fisheries, should be monitored on regular basis. Illegal fishing and over-fishing in the reservoir also should be checked.

Water Quality

Reservoir Water Quality

The pH of the reservoir is in the alkaline range. No significant change in the pH level was observed across the study period.

The reservoir can be categorized as highly eutrophic because of the (i) high EC (>200 μ mhos) (Rawson, 1960; Berg et al., 1958), (ii) high alkalinity values (Phillipose, 1960; Spencer, 1974) and (iii) high bicarbonate (Moss, 1973; Munawar, 1970). As per the criteria of Moyles (1946) the reservoir can be categorized as a hard water body due to the high alkalinity values.

During the monsoon months, across the study period, EC, alkalinity, total solids, suspended solids, sulfates, and chlorides were on rise. This can be due to the contribution of run-off from the catchment, which would bring with it the various forms of solids, chloride, and other soil constitutes. A declining trend had been observed in the case of dissolved solids, and calcium hardness, which can be attributed to the dilution caused by rainfall and run-in.

An increasing trend has been observed during the summer months in the case of total solids, dissolved solids, calcium hardness, and chlorides, which could be explained due to the evapo—transpiration factor. However, the parameters—alkalinity, total hardness, magnesium, and sulfates-showed declining trends.

No significant change was observed in the trends of EC (summer), pH (both summer and monsoons), suspended solids (summer), total hardness (monsoon), and magnesium (monsoon).

Pick-up Dam Water Quality

The pH of the pick-up dam is in alkaline range, similar to that of the reservoir. The water held by the pick-up dam can also be categorized as highly eutrophic because of the high EC and alkalinity.

Across the study period, during the monsoon months, EC, alkalinity, total solids and chlorides were on the rise similar to the reservoir water. This can be attributed to the run-off coming from the catchment which would add various solids, chlorides and other soil constitutes. A declining trend has been observed in the case of turbidity and iron content, which may be due to the dilution factor of both the direct precipitation and the abundant rainwater contributed by the run-off; kindly note the increase in water level holding of the pick-up dam.

The water quality parameters—alkalinity, EC, hardness, chloride, fluoride and total solids—showed an increasing trend during the months of summer. However, the values of turbidity and iron content showed a declining trend.

In the case of pH, during summer, and fluoride, during the monsoon, no significant change was observed.

When the water quality of the reservoir and pick-up dam was compared, it was observed that the pH remained more or less similar; EC, total solids, chlorides were lesser; total hardness and alkalinity were on the slightly higher side.

IMPACTS, DOWNSTREAM

Irrigation

Even though SRP was developed to provide surface water resource for irrigation and reduce the strain on groundwater; it is the groundwater that continues to bear the greater burden. About 58 per cent of the irrigated land is supported by groundwater irrigation while only 25 per cent is under direct canal irrigation Figure—10.2 (Volume 1). The remaining 17 per cent is provided by tanks, which, indeed, had existed even before the SRP was commissioned. These tanks are partly supplied by the canals and partly by rains.

In SLBC command, 50 per cent of the irrigated area is under canal irrigation as compared to 21 per cent in SRBC command Figure—10.2 (Volume 1). This is because of the priority rights of SLBC command over SRBC in the irrigation water supply.

The gross irrigated area from all the sources in SCA is 70.2 per cent. In SRBC and SLBC commands the gross irrigated areas are 67.6 per cent and 72.8 per cent respectively (Figure—10.3) (Volume 1).

The intensity of irrigation in the SCA is 1.33, which implies that the command area is way short of the irrigation facilities for double cropping Table—10.1 (Volume 1). It also points out towards the cultivation of perennial water intensive crops-paddy *(Oryza sativa)* and sugarcane *(Saccharum officinarum)* in the SCA.

Only 36 per cent of the potential created by the way of canal irrigation was actually utilized in the SCA. A mere 21 per cent of the envisaged area is covered in the SRBC command. Even in SLBC, only 50 per cent of the proposed area has been brought under canal irrigation Figure—10.4 (Volume 1). This gulf between the potential created and actual utilization could be attributed to the absence of lining in the canals, inadequate maintenance, and infestation of the canals and tanks by weeds and other vegetations (Chapter 16) (Volume 1).

Problems of seepage, water logging and soil alkalinity were reported in various villages in SCA due to faulty and mismanaged canals and distributaries, especially in the fields adjacent to the canals and tanks.

Making available the irrigation facilities by way of introducing canal irrigation has had a profound bearing on the cropping pattern of the SCA. There is a heavy tilt towards the perennial cash crops of paddy, sugarcane, and groundnut *(Arachis hypogea)*. This has happened even though the SLBC command is supposed to cultivate only the dry irrigated crop of groundnut.

In the absence of proper understanding between the cultivators and project authorities and also amongst the cultivators, occasional clashes were reported with regards to the proper allocation and distribution of water both at the macro and micro-levels.

The two canal systems have bifurcated the SCA not only in terms of supplying the water through separate canals but also by way of introducing conflicts between the cultivators of both the command areas. End-users of both the command areas find the canal water insufficient for the irrigation purposes and blame each other for the shortage.

There is an immediate need to line-up the sub-canals and distributaries and to de-silt the traditional tanks in the villages. A thorough water budget has to be prepared to account for any conveyance and seepage losses. Traditional water harvesting strategies may be implemented, both in the SCA and Sathanur catchment, to augment the water shortage.

There should be a proper framework, and guidelines for the efficient utilization of irrigation water in relation to the cropping pattern and the water demands of the crops.

The governmental as well as non-governmental agencies should generate strong thrust towards involvement, understanding, and participation of the end-users with regard to the water distribution. This may help in the optimal utilization of the water through the canal/tank system. It may also reduce social tensions and inculcate in the farmers a sense of belonging for SRP.

Land Use

The following points emerge from the detailed study of the land use pertaining to SRP:

1. The forest cover in the Sathanur command area (SCA) is a mere 1.1 per cent of the total land under SCA as against the Tamil Nadu State's forest cover of 16.4 per cent and Thiruvannamalai district's (which encompasses most of the study area) forest cover of 24.3 per cent.
2. Barren and uncultivable land constitutes 7.5 per cent of the total reported area (TRA) of 84219.3 ha in SCA, which is double the state's proportion of 3.7 per cent. SRBC command has higher proportion of barren and uncultivable land (8.3%) compared to SLBC command (6.6%) because of the fact that the terrain of the SRBC command region is rocky with hard, stony, and modulus soil (Chapter 12) (Volume 1).

3. Cultivable wasteland forms just 2.3% of the TRA in the study area (Figure—11.11) (Volume 1).

4. Permanent pasture and grazing lands constitute only 0.1 per cent of the land area in the SCA. In the absence of pastures and grazing land, the cattle population has dwindled to one-fourth of what it was 10 years ago in the SCA as per the estimates reached on the basis of survey with the villagers. It was rare to sight the cattle during the field trips.

5. Land under miscellaneous trees and groves constitutes just 0.3 per cent of the TRA in the study area against Tamil Nadu's portion of 1.4 per cent.

6. The fallow lands for the year 1997-98 were reported to be 17.6 per cent of the TRA in SCA. The SLBC command had its 15.5 per cent of land under this category while SRBC command had 21.6 per cent. The reason for the fallow land being more in SRBC command than in SLBC is the presence of coarse loamy soil (Chapter 12) (Volume 1).

7. The net cultivated area in SCA is ~ 60 per cent which is considerably higher than Tamil Nadu's proportion of 42.2 per cent. Following the introduction of canal irrigation there has been a tendency among the farmers, downstream SRP, to bring more and more of the land under cultivation. The total cultivable land in SCA when compared with other basins in the country is far higher: 79 per cent of TRA. For example the total cultivable land in the Tapi basin is 62 per cent and in Krishna basin 61 per cent of the total available land.

8. Efforts should be made to improve land use management in SCA to ease the pressure on the land. The barren and uncultivable lands should be reclaimed and made productive. More of the land should be brought under forest cover, with poly cultures. Villagers should be encouraged to participate in social forestry programs. Cultivable waste lands may be developed as pastures, which would encourage cattle rearing. This in turn would boost the local economy.

9. A comprehensive land use policy need to be developed for SRP based on the requirements of the local community and the state government. In this context, a more intensive micro-level study of the land use/land cover of SRP need to be undertaken using the latest surveillance tools-GIS and remote sensing. This database can be highly useful in evolving a comprehensive land use/land cover policy.

Soil

The extensive study on the soil quality of SRP indicates:

1. The soil in the SLBC and SRBC commands is predominantly fine loamy and coarse loamy respectively.
2. As per the land capability classification, the SLBC command is susceptible to water stagnation due to the poor permeability and erodibility of the topsoil. The lands in SRBC command are even more susceptible to erosion, salinity and alkalinity hazards.
3. The drainage in SLBC command is better than in the SRBC command due to the fine loamy texture of the soil.
4. The crops considered suitable for both the commands are groundnut *(Arachis hypogea)*, millet *(Pennisetum typhoides)*, sunflower *(Helianthus annus)*, maize *(Zea mays)*, sorghum *(Sorghum bicolor)* sugarcane *(Saccharum officinarum)*, banana (Musa paradisiaca), coconut *(Cocos nucifera)*, Chilli *(Capsicum annum)*, pulses and other horticulture crops.
5. The current cropping pattern is tilted towards paddy, groundnut and sugarcane which require intensive irrigation. Only millet and maize are grown as dry crops in the absence of sufficient water for irrigation.
6. Cultivators should be encouraged to take up cropping patterns appropriate for the land-water conditions. Diverse cropping patterns need to be evolved to maintain the soil fertility and optimize the use of water.

7. Technical support may be made available to guide the cultivators in the matters of fertilizer use, pesticide use, irrigation, cropping, and value - addition.

Agriculture

In SCA, the major produce of the agricultural land (78.6%) constitutes paddy, cereals, pulses, spices, vegetables and fruits. Of this, paddy occupies the bulk-37.3 per cent of the total cropped area; sugarcane 17.13 per cent, cereals 18.02 per cent, pulses 4.75 per cent and spices 2.4 per cent follow it.

Among the 22.4 per cent of the area under non-food crops, the bulk (four - fifth) constitutes groundnut. The remaining cropping area (3.9%) is shared by sesamum *(Sesamum indicum)*, castor *(Riccinus communis)*, sunflower, cotton *(Gossypium arborium)* and other crops.

The cropping pattern of the SCA is, thus, dominated by paddy, cereals (millet, maize), sugarcane and groundnut. Of these paddy and sugarcane are perennial crops requiring intensive irrigation. This cropping pattern has emerged as a direct result of the introduction of canal irrigation in SCA, making available plentiful water for long durations. Also, this cropping pattern is governed by short-term socio-economic factors rather than ecological considerations and their long-term socio-economic implications.

To cater to the needs of sugarcane plantations, many sugar mills (one of them being located in SCA) and other agro-based industries dealing with fertilizers, pesticides and other agro products have come up in the vicinity of SCA. Better roads, transportation felicities and marketing facilities have also resulted, contributing significantly to the development of the region.

The intensity of cultivation is only 1.38 in SCA, which indicates that the area has not reached the status of double cropping (Table—13.15) (Volume 1). This could be either due to insufficient irrigation water, which may be the direct result of damaged and unlined canals, or due to the cultivation of water intensive perennial crops like paddy, sugarcane and banana.

The population supported per hectare of the sown area in SCA is 6. Cultivators constitute 40.5 per cent of the work force and 50 per cent of the workers are landless laborers. Among the laborers, 53 per cent are female. This data reflects the poor socio-economic condition of the majority of the populace in the region. In other words, even as SCA has promoted development, the benefits do not yet seem to have reached the majority.

As irrigation constitutes the near-total consumer of the water resources developed by SRP, there is a strong need to reorganize the cropping pattern along scientific lines so that the present over-emphasis on water-intensive crops is reduced and optimal combinations of cropping patterns are evolved which are compatible with the soil and climate of the region.

As half of the work force in SCA comprises of landless labourers, they should be provided with a support system in terms of loans or other viable provisions to earn better livelihood.

Ground Water Quality

1. The pH of ground water in most of the villages in the Sathanur command area (SCA) was above 7. Some villages of Sathanur right bank command area (SRBC) had lower pH of the ground water which varied between 6.5 - 6.9. These villages were Palayanur (6.5), Thenkarimbalur (6.9) and Jambai (6.9) in SLBC and Porasapattu (6.6) and Erudayampattu (6.9) (Figures—14.1 and 14.2) (Volume 1).
2. Ground water samples of Allikondapattu, Thachampattu, Periyakallipadi, Thenmudiyanur and Melandhal had lowest EC (~1500 μ mhos cm^{-1}; Figure—14.3) (Volume 1). In SRBC command, the groundwater of S.Kolathur and Pakkam villages recorded EC values of more than 2000 μ mhos cm^{-1} (Figure—14.4) (Volume 1). These figures indicate salinization of the groundwater.
3. All the samples drawn from the villages of SLBC and SRBC commands, except S.Kolathur, had alkalinity values higher than the permissible limit for drinking water (Figures—14.5 and 14.6) (Volume 1).

4. The groundwater samples drawn from Palayanur, Allikondapattu, Thachampattu, Kandiankuppam, Periyakallipadi, Velayampakkam, Sadakuppam, Thenmudiyanur, Vanapuram, Melandhal and Edathanur in SLBC command had higher hardness concentration than the permissible limit for drinking water (Figure—14.7) (Volume 1).

   In SRBC command, groundwater at Melsiruvallur, Moongilthuraipattu, Vadaponparappi, S.Kolathur, Porasapattu, Erudayampattu, Arambarampattu, Jambodai, Kadavanur, Pakkam, Athiyur, Manarpalayam had hardness concentration in excess to that of the permissible limit for drinking water (Figure—14.8) (Volume 1).

5. In terms of calcium concentration in the groundwater, the villages Palayanur, Allikondapattu, Thachampattu, Kandiankuppam, Periyakallipadi, Devariyarukuppam, Velayampakkam, Valavachanur, Vanapuram, Melandhal, Thenkarimbalur, Jambai and Edathanur in the SLBC command had their ground water exceeding the limits set by Indian standards for drinking water (Figure—14.9) (Volume 1). In SRBC command, all the villages had calcium concentrations more than 75 mg l-1 in their groundwater. The ground water in these villages was, thus, unfit for human consumption (Figure—14.10) (Volume 1).

6. In terms of chloride concentration, the groundwater in Palayanur, Melandhal and Edathanur in SLBC command (Figure—14.11) (Volume 1) and Vadaponparappi, S.Kolathur, Erudayampattu, Kadavanur, Pakkam, Athiyur, Manarpalayam and Kidagudayampattu in SRBC command (Figure—14.12) (Volume 1) is unfit for drinking as per the standards (IS 10500-1993) of drinking water (Table—9.23) (Volume 1).

7. The groundwater from Palayanur, Velayampakkam, Sadakuppam, Valavachanur, Unnamalaipalayam,

Agarampallipattu and Vanapuram villages in SLBC command contained greater nitrate concentration than the limit of 10 mg l^{-1}, thus rendering the water unfit for drinking (Figure—14.15) (Volume 1).

In SRBC command the groundwater samples drawn from Moongilthuraipattu, S.kolathur, Erudayampattu, Kaduvanur and Pakkam contained nitrate concentrations greater than the permissible limit of 10 mg l^{-1} (Figure—14.16) (Volume 1).

8. The phosphate concentration of the groundwater in all the villages surveyed in SLBC and SRBC commands was greater than 0.2 mgl^{-1}, exceeding the limit set by Canadian Department of National Health and Welfare (1969) (Figures—14.17 and 14.18) (Volume 1). The BIS has not yet been set for phosphate concentration in drinking water.

9. Athiyur (SRBC command) had 100 µg l^{-1} chromium in its groundwater, which exceeded the permissible limit for drinking water (Figure—14.26) (Volume 1). Palayanur (SLBC command), Porasapattu, Kadavanur and Arur had higher concentration of zinc in their ground water than specified by the BIS for drinking water (Figure—14.27) (Volume 1). Palayanur, Kadavanur and Arur also had higher concentration of cadmium in their groundwater (Figure—14.28) (Volume 1).

10. Higher concentration of all the major ions, alkalinity, hardness, and EC in the groundwater of SCA could be attributed to the phenomenal growth in agriculture in the region thus leading to the shift towards higher and rather indiscriminate application of fertilizers and pesticides.

11. The groundwater quality of the villages in SCA should be monitored periodically and the villagers should be made aware of the harmful effects of the indiscriminate use of chemicals.

Health

1. Of the 49 villages in the SLBC command, 16 villages were endemic to malaria as evidenced by higher API (Annual Parasite Index) values (Figures—15.1-15.8) (Volume 1).

2. In SRBC command, 8 out of 44 villages had higher proportion of malaria incidence (Figures—15.9-15.12) (Volume 1).

3. As per the 1998-99 health survey, Palayanur PHC in SLBC command recorded the highest API values for elephantiasis; Pavithram, Palayanur and Thenmudiyanur being the endemic villages (Table—15.4) (Volume 1).

4. The SLBC command registered higher malaria incidences compared to the SRBC command (Figure—15.19) (Volume 1). The reasons could be that better availability of water in SLBC command, due to SRP, has favored water-intensive crops. Often, inadequate attention to drainage in these crops has generated pools of stagnant water, favoring the breeding of disease vectors, particularly mosquitoes.

5. Malaria also owes its prevalence in SCA to the poor canal maintenance. Unlined and vegetation-infested water conveyance channels provide lentic conditions favoring the breeding of disease causing vectors.

6. Large scale migration, illiteracy, and poor socio-economic status of the natives of SCA also have a large bearing on the prevalence and propagation of malaria in the region. Majority of the malaria incidences in the SCA are 'imported'—brought to SCA by migrants working in malaria endemic belts outside the study area. Poor nutrition, unhygienic conditions, and careless attitude of the villager's downstream SRP leads to further worsening of the situation.

7. The riverine conditions of the villages—Moongiltturaipattu, Porasapattu, Melandhal and

Vadaponparappi - which are situated on the banks of the river, make them susceptible to the risk of malaria.

8. There is a declining trend in the malaria incidence in most of the villages in SCA during the study period with the exceptions of Alappanur, Vanapuram, Katampoondi, Thenmudiyanur, Valavachanur and Periyampattu in the SLBC command and Rayandapuram, and Poravalur in the SRBC command.

9. The preventive measures adopted by the health officials in checking the menace of water-borne diseases in SCA appear to be adequate as is evident from the declining rate of disease incidence in most of the villages. It appears that largely due to the efforts of public health officials, no outbreak of epidemics has occurred in the SCA, in spite of a large number of factors which had the potential of favoring the propagation of water-borne diseases.

10. Surprisingly, the survey of the villagers' perception in the SCA, indicated that the health facilities are inadequate in terms of poor infrastructure, absence or irregularity of doctors and nurses, and lack of medicines and proper attention.

11. The ratio of PHC (Primary health centres) to population is very low in SCA. Only thirteen PHC's offer health service to a population of over 100,000 (Figure—15.21) (Volume 1).

12. There is an urgent need for proper lining of conveyance channels, better management of irrigation practices, and monitoring of the migrant work-force. Health care facilities also need improvement vis-a-vis establishment of more PHCs and induction of more doctors and medical staff. Some of the alternative systems of medicine, such as homoeopathy, are inexpensive yet they are highly effective. Such alternate systems of medicines may be encouraged so that larger number of people can be covered at lower costs.

13. The agro chemicals—DDT, malathion, temphos, and organo-phosphates—which are used for spraying and fogging to check the growth of mosquitoes and their larvae, pollute the environment with dangerous portents. Efforts are required to educate the villagers on personal hygiene and moderation in pesticide use.

Socio-Economics

The salient findings of the ground-truth surveys and EIA-matrix analyses have been summarized below:

Agriculture

Beneficial Impacts

(a) There has been an overall increase of the area cultivated

(b) The productivity per hectare of the cultivated land has also increased.

Adverse Impacts

(a) The use of traditional bio-fertilizers-cow dung and composted manure- has declined in the SCA.

(b) The farmers are increasingly depending on synthetic fertilizers, pesticides and weedicides.

(c) The menace of weeds and pests in the agricultural fields has been on a rise for the past 10 years or so.

(d) Little attention has been paid towards guiding agricultural practices along scientific lines.

(e) Since the SRP has been commissioned, the agriculture societies and banks have mushroomed in the SCA. However, the lower EIU (environmental impact unit) in SLBC command and negative EIU in SRBC command indicate that the benefits from the agriculture societies and banks have not yet reached the common cultivator.

Irrigation

Beneficial Impacts

(a) The SRP has directly benefited the farmers of SCA vis-à-vis greater access to surface irrigation.

(b) Canal irrigation has helped the cultivators downstream, especially in SLBC command. In the case of SRBC command, the benefits have been diluted due the lower priority rights of the users and faulty canals and tanks (Chapter 10) (Volume 1).

(c) The use of modern implements of agriculture and irrigation by the cultivators has increased after the introduction of canal irrigation.

Adverse Impacts

(a) Traditional flood irrigation method is widely used in the SCA. This mode of irrigation results in the wastage of precious water, which would otherwise be available to the farmers further down-stream. The farmers in SCA should be made aware of the many efficient modes of irrigations such as drip irrigation, sprinklers and lift irrigation.

Government should provide incentives in the form of loans and concessions so as to encourage the farmers to opt for the scientifically proven modes of irrigation.

(b) Low-lying fields in the SLBC command experience problems of water logging.

Canal and Tank Maintenance

Adverse Impacts

(a) During the EIA survey, all the indicators of canal and tank maintenance have garnered negative EIUs.

(b) Due to the paucity of funds, the lining of the sub canals and distributaries is yet to be completed.

(c) The absence of sub-canal lining has facilitated infestation by weeds and other vegetation. This generates indirect adverse impacts such as loss of water flow and public health hazards due to the growth of disease causing vectors.

(d) The silting of tanks and canals has drastically reduced their water carrying capacity. Some canals and tanks have even ceased to exist as no water has flown through them since years!

(e) Seepage from the canals has been reported. Even the lined canals are known to seep, flooding the nearby fields and causing water logging.

(f) There is no optimal water release strategy from the reservoir. The water in the canals is reportedly stopped during the crucial periods of harvesting. The cultivators allege instances of bribery on the part of officials to release water in the canals. There is also a conflict among the cultivators of the two commands with respect to the supply of irrigation water as the SLBC command has priority rights over SRBC (Chapter 10) (Volume 1).

(g) There are no organized channels of communication or cooperation amongst the farmers. This leads to clashes and misunderstandings between them with regard to the distribution of water in the fields. The tail-enders usually suffer as the water is totally consumed by the farmers at the upper reaches of the canals.

(h) The canals and tanks are mutilated and encroached by the cultivators.

Infrastructure

Beneficial Impacts

(a) There is improvement in the income level of an average farmer.

(b) The establishment of primary health centers, which occurred after the commissioning of the SRP, has resulted in improvement of health care in the SCA.

(c) The status of schools and education has become better. Every village has a primary school and about 60 per cent of the villages surveyed have secondary schools.

(d) Basic infrastructure such as roads, power supply and drinking water have improved.

Adverse Impacts

(a) SCA lacks forest cover, which puts tremendous pressure on the fuel wood supply of this region, especially for women.

(b) The employment opportunities are quite less in the SCA as there is only one major agro-based industry (the Kallakurichi Sugar Mill) in the entire region. In the absence of sufficient irrigation water, especially in SRBC command, there is a large-scale migration of the villagers to the cities, defeating the basic purpose of setting-up SRP.

(c) The villagers point out that the infrastructure in several public health centers, and the medical attention paid by the health officials, were inadequate.

The various points that emerge from the present socio-economic study of Sathanur reservoir project are:

1. The widely different EIUs generated indicate divergent views of the project authorities and the end-users. This communication gap between the executives and the end-users must be drastically narrowed if SRP has to benefit the region.

2. The results of inter-parametric correlations based on the perceptions of the end-users show significant correlations (Table—16.8) (Volume 1). This indicates the interdependencies of all the impacts on one hand, and better understanding of the various indicators on the part of an average farmer on the other hand.

3. Farmers in the SCA should be made aware of the scientific approach towards agriculture. The cultivators should be encouraged to adopt efficient irrigation methods and traditional fertilizers such as composted manure, vermi compost.

4. There is an urgent need to introduce drip, furrow and sprinkler irrigation. The traditional practice of flooding the fields should be phased out. The benefits through

the canal and tank irrigation to the SRBC command should be maximized by the efficient use of water and maintenance and upkeep of conveyance channels.

5. Canals and tanks should be de-silted and cleared of the vegetation. Appropriate measures should be taken to reduce the seepage loss. Optimal flow of water from the reservoir and its judicious distribution to the cultivators should be ensured (as per the recommendations in chapter 10) (Volume 1). Instances of bribery to release water in the channels ought to be looked into.

6. As concluded earlier, more of the land should be brought under forests in SCA. Measures ought to be taken to ensure judicious and efficient supply of irrigation water through canals and tanks by way of proper infrastructure and management as discussed in the previous section and in chapter 10 (Volume 1). This will step-up the agriculture production and may check large-scale migration of the cultivators to the nearby cities in search of employment.

7. Monetary support and the necessary infrastructure should be extended to the landless labourers and poorer cultivators.

8. Regular public health surveys and other social services should be organized.

9. Finally, all the departments related to the SRP should be brought under a common scheme of Command Area Development (CAD) that is operational in many other water resources projects in India. This would provide better administration and also better understanding between the authorities and end-users.

Forestry

1. The extensive survey conducted by us indicates that the forest cover in the SCA is very sparse (Chapter 11) (Volume 1). Even the statistics of the government of Tamil Nadu substantiate this.

2. The dominant tree species encountered in most of the villages was *Acacia nilotica*, which has been planted by the Forest Department of Tamil Nadu. According to the concerned departmental sources, *A. nilotica* has been planted because no other tree species would survive the hot and arid conditions prevailing in the SCA. However, the villagers feel that this species is more of a nuisance than a help as it colonizes and encroaches upon their agricultural lands.

3. In the villages of Vanapuram, Varagur, and Arulampadi, the government has planted eucalyptus as part of the social forestry programme. However, the villagers fear that these trees may reduce the water table. The fear is so strong that the villagers want the trees to be uprooted. On their part, the forest officials maintain that they have discontinued eucalyptus plantations and that all the existing trees were planted over 7 years back.

4. During the field trips a few trees such as tamarind, mangoes, neem etc. were found to be thriving in the SCA. The eco-restoration should emphasize multi-species plantations with a good mix of appropriate species based on the geo-climatic conditions prevailing in this region.

Industry

1. There is only one major agro-based industry: Kallakurichi Cooperative Sugar Mill operating in SCA. The mill was established to provide ready market for the sugarcane cultivators in the SCA.

2. The cultivators complain that the mill does not lift the produce on time causing heavy losses to the cultivators.

3. The mill is also implicated in polluting the Ponnaiyar river by releasing its' untreated effluents into it. Ponnaiyar river water is a major source of drinking water for many villages downstream. Fish kills were also reported in the river due to the release of effluents.

The mill authorities deny all these allegations and put the blame on the cultivators. The mill authorities say that out of greed the cultivators plant more sugarcane than the mill can lift. The effluents, they say, are treated appropriately by the effluent treatment plant, which has been installed in the factory premises, and the water is let into a small farm maintained inside the premises. However, the ground-truth survey and the laboratory experiments indicate that the mill does pollute the river significantly.

4. The Ponnaiyar water downstream should be monitored regularly for possible pollution due to the effluents being released into the river by the mill.

5. More of agro-based industries should be established in the SCA. It may also be considered to enhance the capacity of the existing sugar mill so that it can process the entire crop on time.

Water Supply

The Sathanur reservoir meets the domestic water needs of a population of 2,00,000 by supplying 0.4 million litres of water per day. This is a minor but indisputably beneficial impact of SRP.

The Role and Application of GIS (Geographic Information Systems) in the Conservation and Management of Water Resource Projects

In recent times, GIS and remote sensing have emerged as powerful tools in the natural resources assessment. This has become possible due to the recent advancements in the fields of information technology, and communication engineering.

We have developed SRPIS (Sathanur Reservoir Project Information System), a stand alone, upgradeable, GIS package for SRP (Sathanur reservoir project). Apart from its direct relevance to SRP, SRPIS also showcases the development and application of a typical GIS for the environmental monitoring, assessment and conservation of water resource projects.

Sathanur Reservoir Project Information System

SRPIS, Sathanur Reservoir Project Information System, has been designed so as to enable the user to store, analyze, and display spatial information on the Sathanur Reservoir Project (SRP) as maps and attribute data.

The database module of SRPIS has 31 pre-designed forms which enable cognition of a wide variety of data. This module enables the user to manipulate and update the database of SRP on a real time basis. The user can also browse the existing information through *forms*, tables, and *reports*, as one does when using MS Access.

One of the novel features of SRPIS is the EIA sub-module which has been integrated with the software database module of SRPIS. The environmental Impact assessment ratings of the Sathanur reservoir project can be directly fed into this module by various observers: experts, end users, and other groups. The personal details of each observer are stored in the table *EIA_Personaldetails*, and the EIA ratings are stored in a yet another table called *EIA-Matrix*. In the latter, the values are saved cumulatively for each village and for each category of the observer.

The mapper module of SRPIS has been designed especially for GIS requirements. The user can browse through various facets of SRPIS as map layers, thematic features, the attribute data related to these maps, hydraulic designs of Sathanur reservoir, and information on tourist interests in-and-around Sathanur.

SRPIS, because of its powerful features and user-friendliness, would aid in the continuous monitoring and assessment of Sathanur Reservoir Project when implemented on real-time basis.

Bibliography

Abbasi, S.A, Abbasi, N, Bhatia, K.K.S., Khan, F.I. and Sharma, R., 2000. Small Hydro: Potential, Technology and Environmental Impacts, *IPHE*, 2000 (3): 38-48.

Abbasi, S.A., 1989. *Experimental Aquatic Ecology and Water Pollution*, Pondicherry University, Pondicherry.

Abbasi, S.A., 1991. *EIA of Water Resources Projects with Special Reference to Krishna, Mahanadi and Godavari River Basins*, Discovery Publishing House, New Delhi.

Abbasi, S.A., 1997. *Wetlands of India: Ecology and Threats, Vol. II: Asia's Largest Lake (Chilka)*, Discovery Publishing House, New Delhi.

Abbasi, S.A., 1997a. *Wetlands of India: Ecology and Threats, Vol. I: The Ecology and Exploitation of Typical South Indian Wetlands*. Discovery Publishing House, New Delhi.

Abbasi, S.A., 2000. *Environmental Impact of Water Resources Projects*, Discovery Publishing House, New Delhi.

Aronoff S., 1989. *Geographic Information Systems: A Management Perspective*, WDL Publications, Ottawa, Canada.

Aspinall R., and Pearson D., 2000. *Integrated Geographical Assessment of Environmental Condition in Water Catchments: Linking Landscape Ecology, Environmental Modelling and GIS*, Journal of Environmental Management 59: 299 - 319.

Avery T. E., Berline G. L., 1992. *Fundamentals of Remote Sensing and Air-photo Interpretation,* Mac Millan Publishing Company, New York.

Acreman, M. et al 1999. *Managed Flood Release from Reservoirs : A Review of Current Problems and Future Prospects,* Institute of Hydrology, Oxford.

Adibisi, A.A., 1981. Plankton Composition in the Estuaries of the Konkan Coast During the Pre-monsoon Season, *Mahasagar,* 14: 55-60.

Afroz, Ahmad. and Singh, P.P., 1991. Environmental Impact Assessment for Sustainable Development: Chittaurgarh Irrigation Project in Outer Himalayas. *Ambio,* 20(7): 298-302.

Ampofo, J.A. and Zuta., P.C. ,1995. Schistosomiasis in the Weija lake: A Case Study of the Public Health Importance of Man-made Lakes, *Lakes and Reservoirs: Research and Management,* 1: 191-195.

APHA., 1992. *Standard Methods for the Examinations of Water and Waste Water,* Washington DC 20005.

Arora, C.L. et al., 1987. *Indian J. Agric. Sci*: 80-85.

Balakrishnan, M., 1991. Damned Body Blow, *Down to Earth,* CSE, New Delhi, 6(6): 20-21.

Bandhopadhyay, J., 1990. Tehri Dam: Challenge Before the NF Government *Economic and Political Weekly,* 25(5): 243-244.

Bandhopadhyay, J. and Gyawali, D., 1994. Himalayan Water Resources: Ecological and Political Aspects of Management, *Mountain Research and Development,* 14 (1): 1-24.

Bandhyopadhyay, J., 1995. Sustainability of Big Dams in Himalayas, *Economic and Political Weekly,* September 23: 2367-2370.

Bangnold, R.A., 1936. The Movement of Desert Sand, *Proc. Roy Soc.* (London Ser A Nr 892), 157: 594 - 620.

Bath, G.F., 1978. The River Midge Plague, *Karoo Agriculture,* 1: 37-39.

Baxter, R.M. and Glaude, P., 1980. Environmental Effects of Dams and Impoundments in Canada, Experience and Prospects. *Canadian Bulletin of Fisheries and Aquatic Sciences,* pp. 205.

Berg, K., Anderson, Kand Christensen, T., 1985. Furesundessglsar 1950-54. Limnologic Studies Over Fure's Kulkurpavirkning, *Folia Limnol, Scand,* 10: 189-192.

Bernacsek, G.M., 1984. Guidelines for Dam Design and Operation to Optimize Fish Production in Impounded River Basins, *CIFA Tech. pap. 11*, FAO, Rome.

Beveridge, C.M., Ross, G.L., and Kelly, A.L., 1994. Aquaculture and Biodiversity, *Ambio*, 23 (8): 497-502.

Bhatia, K.K.S. Jha, R. Jaiswal, R.K. and Kumar A., 1993. *Sedimentation Problems in Massanjore Reservoir of Mayurakshi River Systems*, West Bengal CS(AR) - 151, NIH, Roorkee, India.

Bhave, G., 1992. Large Reservoirs are Necessary, in National Seminar on Large Reservoir; Environmental Loss or Gain (5-8 Feb.), *Indian Water Resources Society*, Nagpur, India.

Biswas, A.K. and El-Habr, H., 1991. A Holistic Approach to Environmental Assessment of Water Development Projects, *Symposium on Water Resources Management (November 18-20)*, UNEP.

Bodaly, R.A. and Rosenberg, D.M., 1990. Retrospective Analysis of Predictions and Actual Impacts for the Charchill—Nelson Hydroelectric Developments, Northern Manitoba. *In: Joules in the Water Managing the Effects of Hydroelectric Development*, (Deliske, C.E. and Bouchard, M.A. (eds.), Canadian Society of Environmental Biologists; pp. 221-242.

BIS, 1993. *Drinking Water Specifications*, Bureau of Indian Standards.

BIS, 1994. *Quality Tolerance for Fresh Water for Fish Culture*, Bureau of Indian Standards.

BIS, 1993. *Quality Tolerance for Fresh Water for Swimming Pools*, Bureau of Indian Standards.

BIS, 1974. *Irrigation Water Specifications*, Bureau of Indian Standards.

Burrough P.A., and McDonnell R. A., 1998. *Principles of Geographic Information Systems*, Oxford University Press, London.

Burrough P.A., 1986. *Principles of Geographic Information Systems for Natural Resource Assessment*, Oxford University Press, London.

Car, M. and F, C, de Moor., 1984. The Response of Vaal River Drift and Benthos to Simulium (Dip-tera; Nematocera) Control Using Bacillus Thuringiensis Var, Israelensis (H-14), *Onderstepoort Journal of Veterinary Research*, 51:155-160.

Carder, D.S., 1945. Seismic Investigation in the Boulder Dam Area 1940-44, and the Influence of Reservoir Loading on Earthquake Activity. *Bull. Seism.* Soc. Amer., 35:175-192.

Carter J. R., 1989. *On Defining the Geographic Information Systems,* In Fundamentals of Geographical Information Systems: A Compendium, Ripple W. J., (ed.), ASPRS/ACSM, Falls Church, Va., pp. 3-7.

Cernea, M.M., (ed.) 1991. *Involuntary Resettlement Social Research, Policy and Planning in Putting People First: Sociological Variables in Rural Development*, Oxford University Press, New York.

Centre for Science and Environment., 1982. *The State of India's Environment*, A Citizens' Report, New Delhi.

Central Pollution Control Board, 1994. *Basin and sub-basin Inventory of Water Pollution (Tapi Basin)*, ADSORBS 26, New Delhi.

Central Water Commission,1990. *Register of Large Dams in India*, New Delhi.

Central Water Commission, 1991. *Compendium on Silting of Reservoirs in India*, New Delhi.

Central Pollution Control Board, 1993. *Minimum National Standards for Drinking Water*, Government of India.

Chau Kwai-cheong., 1995. The Three Gorges Project of China: Resettlement Prospects and Problems. *Ambio*, 24(2): 98-102.

Chang, J.B., Zhang, G.H., Xu, Y.G., Deng, Z.L., Miao, Z.G. and Song, T.X, 1995. Fish and Fisheries. *In Hydrobiology and Resources Exploitation in Honghu Lake*, Chen, Y, et. al. (eds.). Sciences Press, Beijing, pp. 106-128.

Chari K.B., 2002. *Application of GIS and Remote Sensing in the Environmental Assessment of Two Wetlands of Peninsular India*, Ph.D Thesis, Pondicherry University.

Chari K. B., and Abbasi S. A., 2002, *Kaliveli Wetland: A Typical Bulwark Against Drought*, Proceedings of the Recent Trends in Drought Assessment, Monitoring and Management, IIT Mumbai 115 - 123.

Chari K. B., and Abbasi S. A., 2003. *Primary Productivity of Kaliveli—A Rare Estuarine Wetland of South Indian Peninsula*, Hydrology Journal 26 (1-2): 1-9.

Cole, G.A., 1979. *A Text Book of Limnology*, The ev. Mosley Co., London.

Compasa E.D., Sugumaran R., 2004. *Urban Growth Modelling on the Web: A Decision Support Tool for Community Planners*, Journal Environmental Planning and Management In press.

Cowen D. J., 1988. *GIS versus CAD Versus DBMS: What are the Differences?*, Photogrammetric Engineering and Remote Sensing 54: 1551-4.

Dale P. E. R., Ritchie S. A., Territo B. M., Morris C. D., Muhar A., and Kay B. H., 1998. *An Overview of Remote Sensing and GIS for Surveillance of Mosquito Vector Habitats and Risk Assessment*, Journal of Vector Ecology 23(1): 54-61.

Danahy, J. W. 2001. *Technology for Dynamic Viewing and Peripheral Vision in Landscape Characterization*, Landscape and Urban Planning 54: 125-137.

David, A., Rao, N.G.S. and Rahman, F.A., 1969. Limnology and Fisheries of Tungabhadra Reservoir. *Bull C.I.F.R.I. Barrachpore*, 13:188.

de Moor, F.C. and M,Car., 1986. A Field Evaluation of Bacillus Thuringiensis Var Israelensis as a Biological Control Agent for Simulium Chutteri (Diptera: Nemato-cera) in the Middle Orange River, *Onderstepoort Journal of Veterinary Research*, 53: 43-50.

de Moor, F.C., 1997. Regulated Rivers: Cause of Outbreaks and Solution for the Control of Pest Blackfly, *Lakeline*, 17(3): 22-23, 40-43.

Department of Environment (DoE), 1987. *Handling Geographic Information System*, HMSO, London.

Department of National Health and Welfare 1969. *Canadian Drinking Water Standards and Objectives*, Canada.

Devasundaran, M.P. and Roy, J.C., 1954. A Preliminary Study of the Plankton of the Chilka Lake for the Years 1950-51, Indo Pacif Fish Coun. Bangkok, *In Symposium on Marine and Fresh Water Plankton in the Indo-Pacific.*

Diamond, J. and Case, T.J., 1986. Overview: Introduction, Extinction, Exterminations and Invasions. *In Community Ecology*, Diamond, J and Case, T.J (eds.), Harper and Row, New York, pp. 65-79.

Down to Earth, 1995. Asia and the Pacific, 4(10)

Down to Earth., 1992. Koel Karo Battles On, 1(2)

Down to Earth, 1996. 5(14)

Dutt, R. and Sundharam, K.P.M., 1999. *Indian Economy*, S. Chand and Company, New Delhi.

FAO., 1993. *Aquaculture Production*, 1985 - 1991, FAO Fish, Circ. 815, Rev. 5, FAO, Rome.

FAO, 1997, FAO *Land and Water Bulletin.*

Fernandes, W. and Thukral, E.G. (eds.)., 1989. *Development, Displacement and Rehabilitation*, Indian Social Institute, New Delhi.

Fernandas, W., 1991. Power and Powerlessness: Development Projects and Displacement of Tribals, *Social Action*, 41: 243-270.

Finch J. W., 1997. *Monitoring Small Dams in Semi-arid Regions Using Remote Sensing and GIS*, Journal of Hydrology 195: 335-351.

Fudge, R.J.P. and Bodaly, R.A., 1984. Post-impoundment Winter Sedimentation and Survival of Lake Whitefish (Coregonus Clupeaformis) Eggs in Southern Indian Lake, Manitoba, *Canadian Journal of Fisheries and Aquatic Sciences*, 41: 701-705.

Gaumat, M.M., Rastogi, R. and Mishra, M.M., 1992. Fluoride Level in Shallow Groundwater in Central Part of U.P, *Bhujal News*, 7: 17-19.

Goel, R.S. and Maitra, R.T., 1992. Large Dams, Ecology and Role of Media, *In National Seminar on Large Reservoirs: Environmental Loss or Gain?* Indian Water Resources Society, Nagpur.

Gottschalk, L.C., 1952. Measurement of Sedimentation in Small Reservoirs, T*rans Am Soc.* Civil Engg., 117: 59-71.

Government of India, 1993. Working Group on Development and Welfare of Scheduled Tribes During the Eighth Five-year Plan (1990-1995).

Grant, F.C., 1969. *Report of a Survey of Endemic Diseases in the Environs of the Weija to Assess the Health Implications of This New Lake,* Project Paper; Ghana Water and Sewerage Corporation.

Guha, S.K., and Patil, D.N., 1990. Large Water Reservoir Related Induced Seismicity, *Gerlandds Beitr. Geophysik,* 99: 265-268.

Gupta, M.C., 1992. *Nutrient Dynamics Plankton and Productivity of Reservoir Amarchand, Southern Rajasthan, in Relation to Fisheries Development*, Ph.D Thesis, University of Udaipur, Rajasthan.

Hamilton, A.P and Shedlock, J.K., 1992. *Are Fertilizers and Pesticides in Groundwater? A Case Study of the Delware, Penninsula, Delware, Maryland, Virginia,* U.S. Geological Survey Circular, 1080:16.

Handa, B.K., 1990. Contamination of Groundwater by Phosphates, *Bhujal News*, pp. 24-36.

Handa, B.K., 1983. Effect of Fertilizer on Groundwater Quality in India *In Symposium on Groundwater Development a Perspective for the Year 2000 A.D,* University of Roorkee, India, pp. 451 - 462.

Helgeson, J.O., Stullken, K.E., and Rutledge, A.T., 1994. *Assessment of Non Point Source Contamination of the High Plains Aquifer in South-central Kansas,* 1987, U.S. Geological Survey Water Supply Paper, 2381-C: 51.

Holcik, J., 1991. Fish Introductions in Europe with Particular Reference to its Central and Eastern Part, Lan *J. Fish. Aquatic Science,* 48: 13-23.

Holden, J.M. and Green, J., 1960. The Hydrology of and Plankton of River Skoto, *J. Anim Ecol,* 29: 65-84.

Hrissanthou V., Mylopoulos N., Tolikas D., and Mylopoulos Y., 2003. *Simulation Modelling of Runoff, Groundwater Flow and Sediment Transport Into Kastoria Lake, Greece,* Water Resources Management 17: 223-242.

Huber, W.C., Brezonik, P.L., Heany, J.P., Dickenson, R.E., Preston, S.D., Dwornik, D.S. and Demaio, M.A., 1982. *A Classification of Florida Lakes,* Final Report to the Florida Department of Environmental Regulation, Vol. 1-2, Report ENV: 05-82.

Iyer, M., 1996.The Dam Sham, *Down to Earth,* 4(17): 38-39.

Jain, R., 1994. Mosquitoes Storm the Desert, *Down to Earth,* 3(13): 5-7.

Jain S. K., Goel M. K., 2002. *Assessing the Vulnerability to Soil Erosion of the Ukai Dam Catchments Using Remote Sensing and GIS,* Hydrological Sciences Journal—Journal Des Sciences Hydrologiques 47(1): 31-40.

Jessop, B.M., 1990b. Stock Recruitment Relationships of Alewives and Blue Back Herring Returning to the Mactaquac Dam, Saint John River, New Brunswick, *North American Journal of Fisheries Management,* 10: 19-32.

Jha, L.B. and Jha, M., 1982. Fluoride Pollution in India, *Int J. of Environ. Study,* pp. 225-230.

Jin, X.C., Liu, H, L., Tu, Q, Y., Zhang, Z. S., Zhu, X.A., 1990. Eutrophication of Lakes in China, *Chinese Research Academy of Environmental Science, Beijing.*

Keulegan, G.H., 1938. Laws of Turbulent Flow in Open Channels, *US Natl. Bur Std J. Res,* 21: 707-741.

Khan, A.H. and Miah, S., 1983. The Brahmaputra River Basin Development, *In River Basin Development, Yamen,* M (ch. Ed.,) Tycooly, Dublin.

Khattri, K.N., 1987. Great Earthquake Seismicity Gaps and Potential for Earthquake Disaster Along the Himalayan Plate Boundary, *Technophysics,* 138: 79-92.

Klimas, A and Paukstys, B., 1993. Nitrate Contamination of Ground Water in the Republic of Lithuania, *NGO Bull*, 424: 75-85.

Kothari, S., 1996. Whose Nation? The Displaced as Victims of Development, *Economic and Political Weekly*, June 15, 1996.

Lehmann A., and Lachavanne J. B., 1999. *Changes in the Water Quality of Lake Geneva Indicated by Submerged Macrophytes,* Freshwater Biology 42: 457-66.

Lindberg, K,, 1991. *Policies for Maximizing Nature Tourism's Ecological Benefits,* World Resources Institute, Washington, D.C.

Malavear M. Y., 2002. *The Applications of GIS to Fisheries Science: Recent Trends, Methodological Problems and Challenges,* http:\\www.GISonline.com

Malla, S. and Weiland, N., 1993. Seismic Safety Assessment of Concrete Dams, *Water Nepal,* 3(2-3): 59-70.

Marrsa S.J., Tuck I. D., Atkinsona R.J.A., Stevensona T.D.I., Hall C., 2002. *Position Data Loggers and Logbooks as Tools in Fisheries Research: Results of a Pilot Study and Some Recommendations,* Fisheries Research 58:109-117.

Martin D., 1996. *An Assessment of Surface and Zonal Models of Population,* International Journal of Geographical Information Systems 10: 973 - 89.

Melzer A., 1988. *Die Gewasserbeurteleilung Bayerischer Seen Mit Hilfe Makrophytischer Wasserpflanzen Gefahrdung und Schutz Von Gewassern,* Ulmer V. E., (Ed.), Universital Hohenheim, Stuttgart, 106-116.

Mitra, A., 1995. Development Distress, *Down to Earth,* March 31; pp. 31-35

Moore, M.P., Abeyratne, F., Amarkoon, R. and Farrington, J., 1983. Space and the Generation of Socio-economic Inequality on Srilanka's Irrigation Schemes, *In Marga,* 7(1): 1-32.

Moore, A.M., Brenner, M., Binford, M.W., Deevey, E.S., Ouxk 1988. *Field Notes on a Paleolimnnological Expedition to Five Lakes on the Yunnan Plateau,* J. Yunnan Univ, 10(b): 28-35.

Morse, B.T., Berger, D., Gamble and Brody, H., 1992. Sardar Sarover: *The Report of the Independent Review,* Report Commissioned by the World Bank and Published by Resource Futures International Inc.

Moss, B., 1973. The Influence of Environmental Factors on the Distribution of Freshwater Algae-An Experimental Study II—Role of pH and the Carbon Dioxide-bicarbonate System, *Journal of Ecology,* 61: 157-177.

Moyle, J.B., 1946. Some Indices of Lake Productivity, *Trans. Amer. Fish Soc,* 7: 332-334.

Munawar, M., 1970. Limnological Studies of Fresh Water Ponds of Hyderabad, India-1 The Biotype, *Hydrobiologia,* 35: 127-162

Murthy, C.S., 1997, *Water Resource Engineering—Principles and Practice.* New Age International Publishers, New Delhi.

Mutreja, K.N., 1985. *Applied Hydrology,* Tata McGraw-Hill, New Delhi.

Nikolskil, G.V., 1969. (J,E.S. Bradely, Transl,) Theory of Fish Population Dynamics, as the Biological Background for Rational Exploitation and Management of Fishery Resources, *Edinburg, Oiver and Boud.*

Odei, M.A., 1982. The Prospects of Some Water-borne Diseases in the Area of the Proposed Weija Dam Reservoir Near Accra, *Ghana Journ. Sci.,* 15(2): 219 224.

Odel, M.A., 1982. *Schistosomiasis in Relation to Water Resources Development. Summary Report on a Survey for the Distribution of Vector Snails and the Trematodes of Economic Importance's Which They Transmit in Some Water Bodies in Ghana Lab.Tech,* Rep.100.

Olivera F., and Maidment D., 1999. *Geographic Information Systems (GIS)-based Spatially Distributed Model for Runoff Routing,* Water Resources Research 35(4) 1155-64.

Orland B., 1994. *Visualization Techniques for Incorporation in Forest Planning Geographic Information Systems,* Landscape and Urban Planning 30: 83-97.

Osowski S.L., Swick, Jr. J.D., Carney G.R., Penal H.B., Danielson J.E., and Parrish D.A., 2001. *A Watershed-based Cumulative Risk Impact Analysis: Environmental Vulnerability and Impact Criteria,* Environmental Monitoring and Assessment 66: 159-185.

Ozemoy V. M., Smith D. R., and Sicherman A., 1981. *Evaluating Computerized Geographic Information Systems Using Decision Analysis,* Interfaces 92-98.

Palmer, R.W., 1991. Description of the Larvae of Seven Species of Black Flies (Dip-tera; Simuliidae) from Southern Africa, and a Regional Checklist of the Family, *Journal of the Entomological Society of Southern Africa,* 54:197-219.

Palmer, R.W., 1993. Short-term Impacts of the Formulations of Bacillus Thuringiensis Var. Israelensis de Barjac and the Organophosphate Temephos, Used in Blackfly (Diptera: Simuliidae) Control, on Rheophilic Benthic Macroinvertebrates in the Middle Orange River, South Africa, *South African Journal of Aquatic Sciences,* 19: 14-33.

Pandey, B.N. and Mishra, R.D., 1991. Hydrobiological Features of River Saura, *BIo. J.,* 2 (2): 337-342.

Paranjpye, V., 1990. High Dams on the Narmada: Indian Analysis of the River Valley Projects, New Delhi: Indian National Trust for Art and Cultural Heritage (INTACH).

Parker H. D., 1988. *The Unique Qualities of a Geographic Information System: A Commentary,* Photogrammetric Engineering and Remote Sensing 54: 1547-9.

Pawar, N.J, and Shaikh, I.J., 1995. Nitrate Pollution of Groundwater from Shallow Basaltic Acquifers, Deccan Trap Hydrologic Province, India, *Environmental Geology,* 25: 197-204.

Philipose, M.J., 1960. Freshwater Phytoplankton of Inland Fisheries, *In Proc. Symp, Algol ICAR,* New Delhi.

Ponce, M.V., 1990. *Engineering Hydrology Principles and Practices,* Prentice Hall, Englewoocliffs, New Jersey; pp. 565-568.

Prasad, Y., 1999. Challenges in the Hydropower Development of the Country, In *Challanges in the Management of Water Resources and Environment in the Next Millennium*, Weed for Inter-institute Collaboration, Mazumder, S.K., Kumar Arun., Kansal, M.L., and Mehrotra, R. (eds.) Proceedings of National Workshop (Oct, 8-9), Touchstone, New Delhi.

Purohit, M.V, Sonetha, N.K., and Rathod, K.L., 1992. Two Major Reservoirs of Gujarat and the Environment, *In National Seminar on Large Reservoirs: Environmental Loss or Gain*? (5.8 Feb.), Indian Water Resources Society, Nagpur. India.

Raghuram, N.,1996. Borne Again, *Down to Earth*, 5(3): 28-32.

Rajan, A.P., 1993. *Performance Evaluation of the Run-of-the River Irrigation System—A New Approach*, Ph.D Phesis, Anna University, Chennai, India.

Rajesh, N., 1995. Mekong's Miseries. *Down to Earth*, 3(22): 20-21.

Raju, K.C.C., Kareemuddin, M. and Pradbakar, Rao, P., 1979. *Operation Anantpur, Geological Survey of India*.

Rao, S.N. and Prasad, R.P., 1997. Phosphate Pollution in the Ground Water of Lower Vamsadhara River Basin, *India. Envt. Geol*, 31(1/2).

Rao, M. M., 1997. Mahanadi, Godavari Surplus Rivers: Excerpts. *The Hindu*, Aug. 28.

Rao N.P.P., Kumar, A.M and Chandrasekhar, M.G., 1990. Cropland Inventory in the Command Area of Krishnaraj Sagar Project Using Satellite Data, *Proc. National Symp. Remote Sensing for Aagricultural Application*, New Delhi.

Rao R. Ch., 2000. *The Hindu*, November 11, 2000.

Rast, W. and Holland, M., 1988. Eutrophication of Lakes and Reservoirs: Framework for Making Management Decisions, *Ambio* 17(1): 2-12.

Rawson, D.S., 1960. A Limnological Comparison of Twelve Large Lakes in Northern Saskatebewan, *Limnol. Oceanogr.*, 5: 195-211.

Reddy. J. P., 1998. *A Textbook of Hydrology*, Laxmi Publications (Pvt) Limited, New Delhi.

Reeves, R.R. and Leatherwood, S., 1994. Dams and River Dolphins: Can They Co-exist?, *Ambio,* 23(3): 172-175.

Rubey, W.W. 1933. Settling Velocities of Gravel, Sand and Silt Particles, *Am . J. S.,* 25: 325-328.

Ruggles, C.P. and watt, W.D., 1975. Ecological Changes Due to Hydroelectric Development on the Saint John River, *Journal of the Fisheries Research Board of Canada,* 32: 161-170.

Ruttner, F., 1963. *Fundamentals of Limnology,* 3rd edn. Transl. D.G. Trag and F.E.V. Fry, University of Toronto Press, Ontorio.

Russel, P.F., 1938. Malaria Due to Defective and Untidy Irrigation, A Preliminary Discussion, *S. Mal Inst,* 1(4): 339-349.

Sahai, B., Parihar, J.S., Nayak, S.R., Singh, T.P., Muley, M.V., Diwari, C.B., Tamilarasan, V., Shende, D.M., Samuel, T.V., Thomas, C.V., Gopinathan, G., Vijan, G., Rajmohan,, K. and Nair, G.D., 1985. Land Use Survey of Idukki District, *International Journal of Remote Sensing,* 6: 507-516

Sahu, H., 1992. Tawa Project An Environmental Gain—An Assessment Case Study, *In National Seminar on Large Reservoirs: Environmental Loss or Gain* (5-8 Feb.), Indian Water Resources Society, Nagpur, India.

Salonen, V.P. and Varjo, E., 2000. Gypsum Treatment as a Restoration Method for Sediments of Eutrophied Lakes—Experiments from Southern Finland, *Environmental Geology,* 39 (6): 3-4

Schulz H.K., Smietana P., Schulz R., 2002. *Crayfish Occurrence in Relation to Land-use Properties: Implementation of a Geographic Information System (GIS),* Bull. Fr. Pêche Piscic. 367 : 861-872

Schumanna A.H., Funkeb R., Schultza G.A., 2000. *Application of a Geographic Information System for Conceptual Rainfall-Runoff Modelling,* Journal of Hydrology 240: 45-61.

Sen, A., 2000. Will There Be Any Help for the Poor?, *Time,* May, 2000: 50-51.

Singh, A.N. and Dwivedi, R.S., 1989. Delineation of Salt Affected Soils Through Digital Analysis of Landsat MSS data, *Int. J. Rem. Sens,.* 10 (1): 83-92.

Singh, I.P. et al., 1987. *Indian J. Agric. Econ*, 42: 404 - 409.

Sinnakaudan S.K., Ghani A. A, Ahmad M.S.S., Zakaria N. A., 2003. *Flood Risk Mapping for Pari River Incorporating Sediment Transport*, Environmental Modelling & Software 18: 119-130.

Small, L., 1983. *Irrigation and Human Welfare—A Workshop Report*, International Agricultural and Food Programme, Cook College, Rutgers University, New Jersey.

Smeats, J and Amavis, P., 1981. European Community Directive Relating to Quality of Water Intended for Human Consumption, *Water, Air, Soil Pollut.*, 4; 483-502.

Smith T. R., Menon S., Starr J. L., and Estes J. E., 1987. *Requirements and Principles for the Implementation and Construction of Large-scale Geographic Information Systems*, International Journal of Geographic Information Systems, 1:13-31.

Spencer, D.H.N., 1964. The Macrophytic Vegetation in Fresh Water Lakes, Swamps and Associated Ferns, *In The Vegetation of Scotland*, J.H. Burne (ed.), Oliver and Boyed, Edinburgh and London, pp. 306-425

Sugumaran R., 2002. *Development of an Integrated Range Management Decision Support System*, Computers and Electronics in Agriculture 37:199-205.

Sugumaran R., Meyer J.C., and Davis J., 2004. *A Web-Based Environmental Decision Support System (WEDSS) For Local Government Planning*, Journal of Geographical Systems In Press.

Sunil and Smitha., 1996. Fishing in the Tawa Reservoir-Adivasis Struggle for Livelihood, *Economic and Political Weekly*, April 6: 870-872.

Thakkar, Himanshu., 1995. Damned Despair, *Down to Earth*, 4 (14)L 16.

Tiegland, J., 1999. *Prediction and Realistic: Impact on Tourism and Recreation from Hydropower and Major Road Developments, Impact Assessment and Project Appraisal*, 17(1): 67-76.

Varshney, R.S., 1986. *Engineering Hydrology*, Nemchand and Bros., Roorkee.

Varshney, R.S., 1986. The Politics and Science of Impact of WRPs on Environment, *In Seminar on Environmental Considerations in Planning of WRPs, Roorkee.*

Venkatiratnam, L., Viswanathan, R., Ravishankar, T. and Rao, P.R., 1991. *Major Crops Identification and Estimations of Irrigated Area in Nizamsagar Command Area Using LANDSAT TM and IRS-IA LISS*-1 Data, Project Report, NRSA and APERL, Hyderabad, India.

Venkiteswaran, K., 1997. How Dams Weaken the Oceans. Science and Technology, *The Hindu*, July 7.

Viswanathan, S., 2000. Madras musings; 10(9).

Welcomme, R.L., 1988. International Introduction of Inland Aquatic Species, *FAO Fish*, FAO, Rome, pp. 294.

Whitmore, T.J., Brenner, M., Jiang, Z., Curtis, J., Moore, A.M., Engstrom, D.R. and Wu, Y., 1997. Water Quality and Sediment Geochemistry in Lakes of Yunnan Province, Southern China, *Environmental Geology*, (1).

World Health Organisation, 1993. *The Control of Schistosomiasis*, Second Report of Expert Committee, WHO Tech. Ser 830.

World Health Organization, 1984. Drinking Water Standards. *www.fao,org/doc rep / w4367E W4 367e17.htm-22k*

Xie, P and Chen, Y., 1999. Threats to Biodiversity in Chinese Inland Waters, *Ambio*, 28(8).

Yoder, R., 1981. *Non Agricultural Uses of Irrigation Systems, Past Experience and Implications for Planning and Design*, Draft Prepared for the Agricultural Development Council. Inc., Rural Development Committee, Cornell University, Lthaca, New York.

Zeid, M.A., 1990. Environmental Impact of High Aswan Dam, *In Environmentally Sound Water Management*, Thanh, N.C and Biswas, A.K. (eds), Oxford Univ. Press, New York.

Zhang, G.H., 1991. Study on the Population of the Crucian Carp (*Carassius Auratus* Auratus), In *Honghu Research Group, Institute of Hydrobiology*, Academia Sinica (ed.) Studies on Comprehensive Exploitation of Aquatic Biological Productivity and Improvement of Ecological Environment in Lake Honghu, Beijing, pp. 153-161.

Zhong, Y. and Power, G., 1996. Some Environmental Impacts of Hydroelectric Projects on Fish in Canada, *Environmental Geology*, 14: 285-308.

Zutshi, D.P., Subla, B.A, Khan, M.A. and Wanganeo, A., 1980. Comparative limnology of Nine Lakes of Jammu and Kashmir, Himalayas, *Hydrobiologia*, 72: 102-112